SEA MAMMALS
AND REPTILES
OF THE
PACIFIC COAST

SEA MAMMALS AND REPTILES OF THE PACIFIC COAST

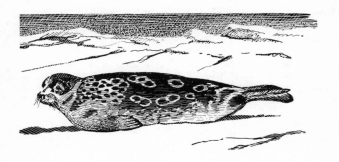

BY VINSON BROWN

MACMILLAN PUBLISHING CO., INC.
NEW YORK

COLLIER MACMILLAN PUBLISHERS
LONDON

Macmillan Publishing Co., Inc.
866 Third Avenue, New York, N.Y. 10022
Collier Macmillan Canada, Ltd.

Library of Congress Cataloging in Publication Data
Brown, Vinson, 1912–
  Sea mammals and reptiles of the Pacific coast.
  Bibliography: p.
  Includes index.
  1. Marine mammals—Pacific coast (North America)
2. Sea turtles—Pacific coast (North America)   3. Mammals
—Pacific coast (North America)   4. Reptiles—Pacific
coast (North America)   I. Title.
QL713.2.B76   1976b   599'.092'63   76-17089
ISBN 0-02-517310-3

First Printing 1976

Designed by Jack Meserole

Printed in the United States of America

This book is dedicated to my second son, Jerrold Vinson Brown, whose time spent in the U.S. Coast Guard off the Pacific Coast taught him a deep appreciation for the magnificent mammals and reptiles that swim in our greatest of oceans.

1920593

ACKNOWLEDGMENTS

I wish to express my grateful appreciation to the following people for help received while I was writing this book: to Mrs. Constance Schrader, formerly a Senior Editor at Macmillan Publishing Company, for her suggesting this subject for me to write and for her encouragement and criticisms of my first writing, which redirected it in a more meaningful way; to Dr. John Sullivan, biologist at the Southern Oregon College in Ashland, Oregon, whose criticisms of my earliest writing made me work carefully for accuracy in the finished work; and to Dr. Robert T. Orr, of the California Academy of Sciences in San Francisco, who kindly answered my questions about some of the scientific nomenclature in the book and helped me use the most up-to-date naming.

CONTENTS

FOREWORD

In describing the sea mammals and reptiles of the Pacific Coast, I have checked numerous scientific sources to add to my own observations and have consulted with scientists when necessary in order to be as accurate as possible. Of course, as most biologists will admit, our present knowledge of most of these animals is very limited and we have a tremendous amount yet to learn. Since one of my main purposes has been to write lively stories about some of the more interesting, important, or well-known species, to help the reader really feel them as living beings, it has been necessary to weave a body of imaginary actions and dramatic events around a framework or skeleton of facts. I feel strongly, from my own observations of these animals and my extensive research, that everything I have written in this book could actually happen as told, but I realize that at this stage of our knowledge some of it may be difficult to prove. Therefore I can only advise the reader that when he wants strictly facts, he should turn to the main descriptions of each animal; but when he reads the stories he should regard them as an honest attempt to catch a mood and a feeling about each animal that would not be possible if we quibbled over every detail. After all, scientists realize that even in a single species there are many variations of behavior, and that some animals, which we might call adventurers or at least trailblazers, can often upset any stereotyped rigid views of how a particular animal should behave.

THE CALL OF
THE SEA

Probably nowhere in the world is there such a fascinating array of sea-loving mammals and reptiles as there is along our Pacific Coast, from Bering Strait in Alaska to Baja California. As more and more people go to sea, or watch these creatures from the beaches, they become aware of this intelligent life that sometimes seems to look back at us human beings from the waves, often paying no more attention to us than if we lived on different planets. The world of the sea is a vast strange world of totally different ways of life from those on land, a world of endless mysteries to be explored. How much more can we enjoy this exploration of the sea and its peoples (even if done in an armchair with a good book) if we really try to get the feel of them, especially the intelligent sea mammals and the calmly deliberate sea turtles.

I remember the first time I went to sea—a rather scared seventeen-year-old, working as an ordinary seaman on the freighter *Golden Wall,* bound for the fabled ports of the Orient. I was a little seasick the first day going out the Golden Gate into a rough swell, and my small sleeping quarters made me feel cramped and lonely. Nothing worked better to get me over both seasickness and loneliness than to stand my watch at the prow of the ship. There the white-crested waves curled and hissed down the steel sides, while every so often some strange denizen of the deep would come up out of the clear green sea. Porpoises or dolphins raced alongside us, breaching the water in marvelous leaps and working in unison like a ballet in the prow waves as they rode them. Then down they went into the sea, their gray-blue shapes leaving twisting trails of bubbles; then up and out again, their entire magnificent bodies flung free from the sea.

More than once a school of blackfish (also called pilot whales), over a hundred strong, surrounded us, their bulbous foreheads making them look like a group of professors, but their dexterity and harmony in the water more exciting than a circus of performing animals on land. Even then I sensed the extraordinary intelligence and vivacity of these creatures, and joined them in their love of the sea. I had no idea at that time of their amazing ways of communication or their understanding of their surroundings, which have been uncovered in recent years by modern scientists.

On six later sea trips to and from Guatemala, San Salvador, Costa Rica and Panama, I was a passenger rather than a seaman, and able to spend even more time at my favorite place at the prow. There I once saw the Southern, or Guadalupe, fur seals cruising out from the islands off Baja California, and porpoising (leaping out of the water) near the ship. Most amazing was the sight once of a great gray whale that had been speared by a swordfish. It tried to shake off its reckless attacker by turning over and over in the sea, slapping the swordfish's eight-foot body onto the waves again and again with a report like that of a cannon.

Then there were the huge sea turtles to watch—green turtles, hawksbills and loggerheads—driving ponderously along through the sea, their immense foreflippers stroking the waves as they moved with the solemn and unflappable dignity of their ancient clans. But perhaps the most delightful of all experiences was to sit on the rocky cliffs of Point Lobos Reserve, just south of Carmel-by-the-Sea, and watch the sea otters playing in the waves and the kelp beds. A mother might be gently suckling her young, grooming it endlessly while she held it on her breast and floated on her back; a raft of sea otter youngsters might be dashing in and out of hidden passageways in the kelp in high-spirited games of hide-and-seek and tag; and a grizzled old male, hoary of head, might be lying luxuriously on his back in the gently rocking sea while he broke open a large clam by hitting it against a rock he held on his stomach for that purpose.

At Point Lobos, near Monterey, and at Point Reyes, on that wondrous peninsula of Marin County in California, I spent even more hours watching both California and Northern (or Steller) sea lions on the rocks, the great bulls fighting for territory, the females nursing their young or barking jealously at other females. Once when I was sleeping alone on the beach they came up around me, their bodies sliding on the sand and their strong scent so close I feared they would crush me. And once a huge bull sea lion chased a couple of boys and myself out of a long dark cave we were exploring, his fearsome roar shocking us into flight.

On the Farallon Islands, thirty miles off the coast from San Francisco, where I went on three rough but wonderfully adventurous visits, I once crept silently to within a few feet of a sea lion colony, to watch their antics and hear their roars and barks. There I had the most terrifying experience of watching a pack of killer whales swim up, their enormous black backfins cutting the waves, and attack the sea lions before some of them could get to the safety of the rocks. The water turned red with blood as the teeth of the twenty- to thirty-foot-long killers sheared through bone and muscle, cutting some sea lions clean in two, swallowing others whole. What screaming there was from the water, what roaring and barking from the rocks, where the sea lions who had escaped helplessly watched their comrades being destroyed.

These and other adventures in observing the sea mammals and reptiles of the Pacific Coast led me to want to write a book about them, a book that would take the reader into their intimate lives and enable him to grasp their feelings and their intelligence, their strange ways of living, such social systems as they have, and their sometimes magnificent and tragic rivalries. A century or so ago, naturalists wrote of the humanlike qualities of the sea mammals they had observed; later scientists have revolted against those ideas and called them anthropomorphic fantasies. I believe both extremes were mistaken, and am glad that the pendulum of scientific opinion is now swinging toward the middle ground where we can appreciate those humanlike feelings and flashes of intelligence among sea mammals while remaining aware of their great differences from us.

For this book I have selected some of the more interesting or readily seen animals for stories that enter their lives in some depth, so the reader can begin to understand the uniqueness and wonder of these lives. To all my descriptions I have tried to add enough of special interest to make each stand out distinctively as a being worthy of investigation and knowledge. By understanding them, the reader too will find himself answering and enjoying the mysterious "call of the sea."

ORDER

CARNIVORA

(Carnivores)

This book starts with the carnivores in its descriptions of species because it is this great order of flesh-eating mammals that has contributed probably the two most recent sea mammals either to leave or begin to leave the land and go down to the sea. With the polar bear in particular we can catch a glimpse of what it is that has caused so many mammals in the past to give up their land life and return to Mother Sea, where all life began.

The order Carnivora is so diverse both in the appearance and the characteristics of its different divisions that it is hard to characterize it as a whole. All Carnivora have three things in common: a set of long sharp canine teeth for seizing and killing prey; three incisor teeth on each side of both jaws; and a "divided" front foot, which may have four toes, but in some cases, five. Most carnivores are fairly strict meat-eaters; but some, such as the bears, the raccoons, and the skunks, are omnivorous and eat a wide range of foods, including many plants.

There are only two species of marine Carnivora in our area: the polar bear (family Ursidae), which is only partly marine, as it does some swimming to catch food, but spends most of its time on ice or land; the sea otter, a member of the weasel family (Mustelidae) that has become almost fully seagoing. The characteristics of each will be described in the following pages.

URSIDAE:

Bears

Body thick and heavy, tail very short, claws, especially those on front feet, developed for digging in ground or breaking open rotten logs and so forth.

Polar Bear

[Ursus maritimus]

Eskimo name: Nanook-Soak

Description   Male with shoulder height up to four and a half feet or more; length seven to ten feet or more; weight up to 1,250 pounds or more, but rarely up to a ton. Female, shoulder height up to four feet, length six to eight feet or more; weight up to 700 pounds or more. Neck much longer, head slimmer and shoulders less heavy-looking than in black or grizzly bears. Both sexes of polar bears are generally colored white with yellowish overtones, though some more yellowish or grayish ones are found. The young usually appear whiter than the adults, but in all the nose and eyes stand out black. Hair grows on the soles of the feet to give better footing on ice and to help in swimming.

Range and habitats   Ranges over most of the Arctic areas wherever there is pack ice and particularly where seals are found, coming as far south in the Bering Sea as the Pribilof Islands, and very rarely, if the pack ice or icebergs drift that far, to the Alaska Peninsula. So it is generally found north of fifty-six degrees north latitude. Though it prefers living and hunting on the pack ice, especially where there are open leads caused by winds or currents, and where seals come up for air, the polar bear also travels inland, especially in summer, to eat carrion and much vegetable matter.

Locomotion, food, and length of life   Though mainly a land mammal, travelling quietly over ice and snow, the polar bear is a good swimmer and has even been found swimming a hundred miles out at sea. Its speed is probably no greater than about two to three miles an hour, so it rarely can catch a seal by swimming but must use stalking for this purpose. Food consists largely of seals and an occasional small whale in fall, winter and spring, but in summer it eats mostly carrion and plant food. Females become adult and are able to bear cubs at about three to four years; males are adult at about five years and rarely live to more than twenty-five years old, the maximum life-span being about thirty-four years.

A Life Story

PRELUDE

Nanook, the great white sea bear of the Far North, is one of the most mysterious and remarkable animals on earth. It is in great need of protection, for the men who hunt it for its skin or as a trophy of the chase have exterminated it ruthlessly over wide areas. Only the vast regions of the uninhabited ice fields and icebergs, rugged and difficult for men to penetrate, have kept it from being

Polar Bear [*Ursus maritimus*]

wiped out completely. Nowadays seaplanes can roar into the farthest wilderness with their hunters, making life for Nanook more and more difficult; for against the high-powered rifle he has no defense except to hide.

In the old days he was a king of the north, feared by every living thing he encountered, except the bull walrus, the killer whale, and an occasional brave Eskimo especially dexterous with harpoon or knife. But the Eskimos of the old days killed a bear only in self-defense or because they needed his warm fur and his meat. To some Eskimos he was their "medicine bear," and they dared neither kill nor eat him. In those days, let the polar bear see anything moving on the ice or snow and he considered it possible food. Moving with the quietness of a shadow, invisible as a white ghost which merged with ice and snow, he would approach, slowly and patiently, until near enough to spring. Whoever then heard his rush and roar, whoever saw his great teeth like white knives, knew a terrible fate was upon them!

Since we are speaking about polar bears mainly as sea mammals in this book, it is well to consider how their ancestors may originally have become seekers of water pathways. We know that many land mammals turned back ages ago to the sea, particularly the seals and the whales, the latter becoming so completely adapted to water life that men for centuries called them "fish"! This difference in sea-adaptation suggests that probably at two different and widely separated periods, the whales first and the seals second, the ancestors of these modern marine mammals found life on land too hard. Possibly a great world-wide drouth, or the sinking of land masses beneath the sea, gradually forced those best adapted to the water to find a way through natural selection to come back to Mother Sea.

But we also know that the polar bears were not among these more ancient returners to the sea. Their adaptations to ocean life are not nearly as complete and are very probably much more recent—probably within the last million years. This is short indeed compared to the twenty-five million years or so of the seals, and the approximately forty million years since the first fossil we can locate of the seagoing ancestors of the whales. The polar bears appear to have begun their gradual change to a partially seagoing life during the hundreds of thousands of years of the great glacial periods, when the great ice may have gradually cut them off from the land animals they were used to preying on and introduced them to the seals as a new and better source of food, for by then the seals had become adapted to and were numerous in the ice-bound northern seas. If so, it was a case of a new and untouched niche of life, ready for exploiting, into which animals driven by hunger naturally found their way and their needed prey.

Going out to sea on the ice to hunt, polar bears in those ancient days were sometimes forced when the ice broke up to swim for their lives, and, in the excitement of the chase, some may have dived into the water to try to catch their escaping prey. So, by natural selection down through the centuries, those bears survived in this far northern icy-cold environment whose adaptations could meet the challenge: extremely thick warm fur, impenetrable by either water or cold; broadened hindfeet, partly webbed and with fringing hairs to help in swimming; and a streamlined figure with less resistance to the water than is found in other bears. The polar bears today are possibly turning gradually more and more into sea mammals. If so, they are only at the beginning of a very long journey.

NANOOK, KING OF THE NORTH

How huge and powerful are the polar bears, but how tiny and helpless were their beginnings! Born in what the

Eskimos call an "iglooviuk," a den hollowed out under layers of snow and ice by his mother, Nanook lay like a squirming red naked worm on her warm belly and thighs, seeking blindly for the delicious milk she gave him, unaware of his future as an ice lord and heroic hunter. He also did not know his future as an "ah-tik-tok," as the Eskimos say: one of "those who go down to the sea."

So, as the milk filled out his body and he began to grow toward his destiny, he showed resemblances to his sea-seeking ancestors. His body became gradually covered with white hair or fur that had an odd fluffiness about it, unusual in mammals. It is hair especially created to withstand the cold of icy water, for it traps air bubbles within it. Also, when the bear leaps out of the water onto the ice and shakes his body, the water is sprayed all around him, and the hair itself becomes almost instantly dry. If you looked at his feet, especially his hindfeet, you would see that they were becoming splayed and were also fringed and padded with special hairs that would make them act both as paddles in the water and a sure footing on slippery ice.

The young cubs do not leave the warm iglooviuk and come out into the weak Arctic sunlight of mid-April spring until they are ready to run and scamper about over the ice and snow. When their mother orders, they must also be ready to dive with her into an open lead of water between two ice floes, and swim behind her like two tails to a dog, or grab onto her tail and be pulled. Most polar bear mothers have two cubs, though some new mothers have only one. A rare few get the task of caring for three. The more common number of two ensures that a cub has someone to play with, to develop his muscles, and learn to time his movements for future hunting.

Down icebergs and snowbanks they often toboggan, over and over, gleefully rolling in the snow, building up a speed that will be of value when, as mature bears, they wish to close quickly with their prey. Small growls and roars accompany their mock fighting, though sometimes it gets serious and furious enough so that their mother reaches out a powerful paw and knocks them apart.

As they grow older, however, their primary purpose in life is to watch their mother swimming, diving, climbing, hunting, hiding from enemies and avoiding the force of storms; they absorb everything they see over and over and finally begin to put it into practice. Their knowledge comes to them through many avenues, not only through the eyes, but the ears, the nose especially, the taste of the tongue, and the touch of paw or nose. And hidden from us, but definitely present, is a strange special knowledge that may come from several sources and which we haphazardly call the sixth sense. Nowadays Nanook needs this last sense most of all, for it may tell him when man is near and that it is time to hide from the greatest killer of all!

His nose is the most obvious help to the polar bear in learning about the world around him. He is always standing up on his hind legs swinging his head and nose around in circles like a snake, testing the air, especially from upwind. Nanook's mother was constantly testing for the smell she specially liked—seal, to hunt—and for the three smells she did not like—men, wolves and male polar bears; for, at this time of year, the latter sometimes kill cubs if they can catch them.

In the late spring of the Arctic, the leads of open water between ice floes open up so that ribbon seals, harbor seals, bearded seals, and other seals can haul up on the ice like great dark slugs to rest, snooze, and scratch in the weak sunlight. It was at this time of year that the mother bear showed her cubs how she and her kind can become partly sea mammals, using both ice floes and open water to sneak up on their prey. Indeed, like a snake, she would wriggle over the ice, moving only when the seal's head was down in sleep, but lying perfectly still when the head was up and watching for enemies, her whitish fur blending with ice and snow and her con- spicuous black nose hidden by her paw.

The cubs would watch her until she came to the dark water of an open lead between herself and the seal. Then she would sink ever so softly and slowly into the water, swimming underwater across the lead and up to the edge of the farther ice. Then she would come up with a leap onto the ice that would send her sliding toward the seal. Now her exquisite sense of timing told her just when to leap, and where and how to aim her great white body in order to meet the seal at the point of a great V as it threw itself forward in a strange looping motion toward the safety of the sea. Out would shoot her paws, armed with several sickle-shaped claws that clamped upon the seal's body and drew it close so that the huge canine teeth could strike for the back of the head or neck, bringing almost instant death, while blood gushed from the seal's nose onto the ice.

Every move their mother made was watched intently by the cubs, who drank into their inner beings every hunting strategy of their kind. Soon they would be trying their own ways to sneak up on a seal; but it would be a long time yet before they felt the satisfactory snap of their jaws and teeth in a live seal's flesh, though they would eat many seals their mother caught for them.

During the summer most polar bears of the Alaska region move to the mainland or float on large ice floes or icebergs to the islands in the Bering Sea. Here they be- come for a while more omnivorous, like most of the bears to the south, feeding not only on carrion, bodies of dead animals cast up on the beaches, but also on grass and other plants in the Arctic meadows. At this time they are mainly land animals, so let us return to the sea.

By the fall the cubs weighed well over a hundred pounds each and were big enough and experienced enough to catch crabs on the rocks and in the tide pools of the sea's edge, and ratlike lemmings on the edge of the land. Then the snow and ice began to close in and the sea wind blew cold from the North Pole. After a rush of seal-hunting, while there were still plenty of open leads and seals resting on the ice, the cubs' bellies were so full that they grew roly-poly with fat, and their mother knew it was time to lead them inland to hole up in a bigger igloo-viuk and dream away the long night of blizzards and the deep freeze of midwinter. It was in the spring, as year-lings, that they began to learn more deeply the mystery and terror of the sea.

Polar bears are as completely in control of their en-vironment as is possible, as long as they are on the top-side of the ice and as long as a storm is not approaching. Some combination of the sixth sense and the sense of smell, perhaps a tingling along the nerves of the back felt from the change of atmospheric pressures, told the mother bear that a storm was on the way that April in their direction, and that somewhere far off the great sheets of ice were moving and buckling, under the tremendous pressure of rising and falling water. She started the cubs running away from the danger, hoping they could reach the solid rocks of an island, or perhaps a very large iceberg, before it was too late.

A polar bear runs about twenty to twenty-five miles an hour when frightened or angry, but cannot keep this speed up as long as a dog or a wolf. Only an occasional savage snarl of warning kept the cubs running after their mother as fast as she would like, but gradually they picked up in the air the terrible feeling of something coming, some-thing beyond the control of any bear, no matter how big! When the ice began to move under them, they whined with fear and poured forth their last strength. There was a terrible roaring, rending, cannon-exploding sound behind them, as the sea ice beneath them began to rise with a great jerk that nearly sent them sprawling. But with their last strength the three bears hurled themselves clawing up the steep side of a big berg, to a platform from where they could look back and see the ice overturning upon itself below them, as ten-ton sheets were pushed up and pitched over each other with the rending and explosive sound of a raging artillery duel. Trembling, they moved as far away as they could from the edge of the platform, to a place where the now blizzard-driven snow would cover them and give them warmth in which to sleep.

It is the frequency of these sounds in the Arctic, so like gunfire, that explains why a polar bear will sometimes seem to stand stupidly on what he assumes is the safety of a rock or a berg, wondering what is hitting him, while a man with a high-powered rifle pours bullets into his hide. Death comes, and there is no sportsmanship and no understanding on either side.

Looking at the leads of black water between the ice floes in late spring, the cubs might have wondered why their mother sometimes led them into the water to swim across to the other side, but at other times took them completely around a lead on the solid ice. She would stand by such a lead, sniffing, watching, feeling, stretching all her senses as if she were sending her very being down into the dark waters to find what was there. Once they saw the great black sail-like dorsal fins of killer whales cutting the waters of a wide lead, and knew from the way their mother growled deep in her throat and backed them away from the water that these were terribly dangerous creatures.

One morning they saw a giant male bear come up over one part of an iceberg and slide down the farther side into the black water of an open lead. He had done it possibly in a moment of play and fun, but suddenly there was no fun anymore and, huge and fierce as he was, he was swimming for his life! A walrus bull had surfaced thirty feet from him, lifted its nearly two-foot-long tusks out of the water in an ominous gesture, and turned toward him with gleaming eyes. Both bear and walrus knew the bear had little chance against a full-grown walrus bull in the water, for the walrus's body and fins drove him through the water with twice the speed of the bear. The walrus dove underwater and the bear screamed as the walrus rose beside him in a moment, bursting through the surface, tusks rising and falling in one fierce sweep. The bear twisted his body desperately, so the tusks only raked his side, rather than penetrating deeply. Blood streaming from him, the great bear dove, came up again by an ice floe, and flung himself up onto the ice. But again he screamed, for the walrus came like an avenging demon out of the water behind him, tusks poised like spears for the fatal thrust.

Now the bear rolled sideways with all the desperation of utter fear, his fierceness and his courage gone. Into the face of the ice the tusks drove like hammered spikes, missing by half an inch! The walrus jerked them out with all his twenty-five hundred pounds of weight and strength, and struck again, but the bear had fled and was safe away. No wonder the cubs looked on the great walrus bull with deep fear after such a demonstration of deadly power. They knew he and the killer whales were their ultimate enemies in the dark waters.

When another winter had passed and another spring followed the great sleep, the cubs were no longer cubs but 500- to 600-pound bears, and their mother drove them off, in a burst of snarling fury, to find food on their own. Now she could seek a new mate, while they wandered alone in the strange way of polar bears. Nanook, more than his sister, faced a period of great danger, for he was still clumsy and new at hunting, and a big male polar bear could run him down and kill him, if he so chose.

To avoid this Nanook wandered far out on the floes in Bering Sea, seeking and staying with large icebergs wherever he could find them. He knew that, with his lighter weight and sharper claws, he could crawl up their steep slopes faster than a larger male, and so escape.

He was into his fourth winter when the urge to hole

up in an iglooviuk ceased and, like most male bears, he spent the cold months roaming the ice, instead of taking the long sleep. A severe blizzard would of course drive him into the shelter of some berg, where he would dig a hole in a snowbank and let the snow form a blanket over him, while his warm breath kept a hole out to the air.

He had to carry out very carefully the correct way of hunting seals. When the ice is so solid everywhere the seals' only way of staying alive is by gnawing blowholes to the surface and working constantly to keep them open, so they can come up to breathe. The polar bear has to locate these holes either by scent or sound or by both, the sound taking the form of a snort or series of snorts as the seal blows through the hole in the ice, and the strong seal smell is wafted on the wind. Two kinds of seals are commonly found making these holes through the deep pack ice, the smaller ribbon seal, weighing from 100 to 200 pounds, and the much larger bearded seal, weighing from 700 to 900 pounds. The larger seal is stronger and can keep open more holes, making it harder to catch him. Seals are helped to disguise their holes when heavy snow falls on them, as snow conceals both the smell and the blowing noise.

The Eskimos say that Nanook sends his spirit down into the sea on the feet of his breath to find out where a seal is making his holes and just when he will rise up through the dark waters. There is no doubt among Eskimos that Nanook has keenness of sense beyond our knowledge, he is such a successful hunter of seals.

Over the ice Nanook walked quietly, under the glowing and flashing colors, the long, streaming, shimmering curtains of the Northern Lights that made the ice and snow pulsate and glow. His entire attention was focussed on listening, smelling and feeling; and in his feeling perhaps lay the truth of the Eskimos' theory about him. So in time he sensed or smelled or heard the presence of his first seal blowhole, the hole of a ribbon seal. He dug down through the snow, and with his claws opened the blowhole wide enough to get his paw down easily. He worked as quietly as possible, for seals have sharp ears and will stay away if they hear noises near one of their holes.

Sometimes a pair of bears, perhaps two who have grown up from the same litter, will cooperate to hunt a seal at its hole. One waits quietly by the seal's blowhole, while the other walks noisily away to make the listening seal below the ice think it is now safe to go to its hole. It is intelligence and learning ability of this kind on the part of polar bears that proves that their reputed stupidity—as when they stand still when shot at—is not stupidity at all; rather, it is a failure to evolve in a few generations responses to such a completely strange thing as a high-powered rifle, especially since the explosion of the rifle sounds exactly like breaking ice.

Nanook has now either to wait quietly by his dis-covered hole until the seal comes or to locate several holes and try to guess which one the seal will use next. Tremendous patience is needed, and the bear holds per-fectly still, sometimes for hours. The Eskimo hunter often imitates his patience, standing over a seal's hole for hours, harpoon poised, and praying perhaps to the spirit of Nanook to help him in the hunt.

The seal itself is wary. If it comes to a hole and catches the faintest whiff of bear or man, it leaves immediately. Nanook has rubbed himself with snow to disguise the smell, and waits and waits. At last he becomes aware of a shadow rising beneath him under the ice, and with it a whisper of sound. His paw is poised, his whole body tensed. When the seal's nose rises under the ice and a gasp of hot and fetid breath explodes up through the hole, Nanook's great paw comes down like a pile driver, but sideways, so that the spread claws can hook and catch the seal by the neck. Now there is a new explosion, the seal's body whipping strenuously to get away, while the bear's tremendous and splendid strength exerts itself to tear the seal up through the hole in the ice. In a few seconds Nanook's prey lies before him, neck broken, blood spurting from the nose, the dark eyes glazing. One animal has found a new lease on life through the death of another.

The hunting of the bearded seal Nanook finds more difficult. Its great size and strength, its more numerous blowholes, and perhaps its higher intelligence make it an elusive and difficult prey. Several times he has found the opening of the blowhole too small, and can't get enough of his forearm down through the ice to use his full strength; consequently, the bearded seal got away. At other times the seal uncannily detected his silent presence above the ice and did not come close enough to be caught. But hunger made Nanook both desperate and wise. Aided by the heavy smell of a bearded seal, he found three blowholes not too far apart, and piled up snow to form a high triangular trail between the holes to hide his shadow under the ice. Then he dug out the top of each hole so thoroughly he could get in both front legs and part of the shoulder. In doing this he so impregnated his front fur with the seal's odor that his own was masked. Then with his usual patience, he settled down to wait and watch.

After an hour and a half of waiting, his ears caught a whisper from the silence under the ice, and up out of the black cold of the depths loomed the big seal's head, the nostrils flaring to snort out the used air of its long underwater swim. At that instant Nanook struck, his paw shooting down from the quickly extended arm and the curved, needle-sharp claws sinking into the thick blubber of the neck with a death grip. He was almost jerked bodily through the ice, for it was a battle of seven hundred pounds against seven hundred pounds, and the balance between the two powerful beasts was delicate. The seal had the advantage of gravity to pull downward; but the bear had the advantage, if he could bring them into action, of his four huge, daggerlike canine teeth.

After the first great flurry, in which he was shaken back and forth over the ice, Nanook threw all his strength into an incredible upward heave that brought the seal's whiskered head within range, and sank his teeth deep through muscle and bone, tearing until the blood gushed forth. The struggle of the seal grew less, and the bear, his body almost half down through the ice, could pull himself higher and drag the body on top of the floe.

After eating and resting, he dragged the remains to a deep snowbank, piled high by blizzard winds, and dug a cave for his prey. The fifty-degrees-below-zero cold would keep the meat fresh for a long time. Replete with the good flesh he had consumed, he lay in his cave and slept. Many hours later a sound dragged him up from his sleep, and he snarled savagely as he saw an Arctic fox tugging at a piece of meat from the farther end of the seal's carcass. He struck, but too late—the snow-white fox fled with a piece of red flesh in its jaws. These scavengers of the great ice follow polar bears wherever they hunt, except in water, as the jackal follows the tiger, waiting and watching with infinite patience for the chance to steal a meal.

Three great, lean Arctic wolves came next, apparently having strayed out onto the solid ice of the winter pack, their red tongues hanging out and drooling as they smelled the fresh kill. Each weighed over a hundred pounds and had jaws strong enough to crack an inch-thick bone, but they heeded the deep growls of Nanook and stayed away from his prey. Much later, eager to hunt again, Nanook stalked off proudly, leaving them the few bones that he had not eaten.

April is the season of heat for the female polar bears, either three-year-olds who have not yet mated, or older females who have got rid of their two-year-old cubs. The three-year-olds, if mated, will probably give birth to only one cub the first time, but the older mothers generally have two, more rarely three. They come out of their igloo-viuk dens thin, hungry and waspish of temper, until some seal meat or other flesh has calmed their nerves. Then they are ready to be mated.

Nanook in his fourth year found a young female with whom he played and romped over the ice packs, until she at last submitted to mating. This had no sooner happened than a bigger bear, with a horrendous roar and a great showing of teeth, drove Nanook away from her. It was not until his sixth year that, weighing a full thousand pounds and glorious in full strength, Nanook did not draw back, but met the usurper tooth to tooth and claw to claw. Most of the fighting was done by tooth, rarely paw, and Nanook had to twist his long sinuous neck constantly to keep from being grabbed and shaken. Snarls and roars shook the air for a mile around while the female crouched and watched the two great bears standing up to each other, their necks red with blood. At last Nanook saw an opening and struck and clamped his teeth on the neck just back of the head of his opponent where he hung on, snarling. The other bear, suddenly anxious to get away, hauled back and shook himself till he could tear loose and run off, leaving a bloody trail. Forgetful of his own wounds, Nanook turned to seek the female.

The great male polar bear is essentially a loner, staying with the female only during the mating season. He wanders as a nomad over the ice sheets and icebergs, sometimes onto islands or the mainland, especially during the summer months. He is occasionally found swimming far out at sea, vulnerable to the walrus and killer whale, but driven by necessity to seek a new ice pack or island on which to land. Perhaps he has drifted south with some ice island and at last has been carried too far from land or other ice and must find his way back. He is partly of the sea himself, but still partially alien, not a complete sea mammal yet.

It was in the darkness of the long Arctic night—in the freezing cold of the great ice sheets, in the time of few leads of black open water, in the time of the great hunger over the north—that Nanook had a chance to get even with the walrus and the whale peoples. The killer whales had long ago retreated far to the south, for their back fins are too high to be used without danger among the ice packs, and they also move where food is more plentiful. But sometimes a pod or herd of walrus, or of the small white whales, or belugas, or of the still smaller narwhals, with their single strange unicornlike tusks, gets trapped in a narrow black lead of water between the ice, far from open water. Forgetful of the danger, they stay too long in the north, and soon it is too late to go south. Then it becomes a struggle against death, because the lead or at least some blowholes have to be kept open for air for the whales, and the walruses have to be able to dive for clams. It is then that they are most vulnerable.

One time Nanook did what polar bears have been seen doing in the London Zoo, and that Eskimos have reported seeing in the Arctic. He seized chunks of ice from an iceberg and threw them around, batting them with his great paws as kittens do with balls of twine. By accident or intent he threw a big chunk down onto a herd of walruses at a narrow lead blow, killing a cow that he could feed upon when the herd later sounded into the water to seek food. Another time, at a similar place, where walruses were trapped by a single lead, he sneaked up on a sleeping bull, weak from too little food, and seized it by the neck from behind, holding on against all its desperate struggles. He finally could chew through to the spine and the bull collapsed on the ice, deep in blood.

Perhaps Nanook's greatest triumph as an ice nomad was to find a pod of six white whales trapped in a small lead near the Bering Strait and drag them out one by one on the ice to be killed, though each was around a ton in weight. That winter he himself weighed close to three-quarters of a ton, a true monster of the north, splendid in muscle and power, owner of enough frozen whale meat to feed himself, a wolf pack, and two dozen foxes or more until spring came!

MUSTELIDAE:

Weasels

Sea Otter

[Enhydra lutris]

Description Length of males five to six feet, weight sixty to one hundred pounds; females four to five feet, weight thirty-five to sixty pounds; color, dark brown to black, usually frosted with white hairs on the head and upper neck. Pups have woolly yellow brown hair. Hindfeet webbed and hind legs especially flattened to take form of flippers for better swimming; front feet not webbed, but furnished with retractile (catlike) claws for seizing food. Tail rather short and thick but also flattened, little longer than legs; used as rudder or oar. The fine inch-thick fur traps bubbles of air when the otter swims, and protects it from the cold and wet. However, otter fur is especially vulnerable to tanker-spilled oil, which may destroy the insulating value and bring quick death from cold and chill. When the pup loses its yellowish brown fur, it appears in adultlike fur, but is called a cub until it reaches full size. An unusual feature is a fold of loose skin across the chest used by the otter as a pouch in which to store or carry food found underwater.

Range and habitat The sea otter's range at one time extended from northern Japan through the Aleutian Islands of Alaska and down the Pacific Coast of North America to the middle coast of Baja California. At present the range is much more spotty, with plentiful sea otters among the Aleutian Islands, somewhat more than a thousand along the California coast from the Santa Barbara Islands to Sonoma County, and transplanted colonies increasing in number along the Oregon, Washington and British Columbia coasts. They are absent from or rare in the Puget Sound region and those reported at San Juan Island near northern Washington proved to be river otters who had taken to the sea in that area. River otters have much longer tails and never float on their backs, as do the sea otters. Sea otters prefer living in or near large kelp beds, but are also found off rocky shores without such beds.

Food and locomotion   Though the main food appears to be shellfish (especially abalones, mussels and clams) and sea urchins, some fish are caught by the more active otters, especially the slower-moving bottom fish. The teeth used in chewing this food are distinctively different from those of any other animals of the order Carnivora. There are no sharp cutting edges on any of them, even the canines, and the postcanines are wide and powerfully built for crushing shells. The lower incisors curve outward and are used to scrape flesh out of clams and abalones. The sharp curved retractile claws of the front feet, so much like those of a cat, are used when the animal is swimming along the bottom or near a rocky ledge, to seize and pull seashells, urchins and other food from the rocks.

The flipperlike hindfeet and legs are used as powerful propellers, while the flattened tail both steers and is sometimes waved from side to side like a sculling oar. Since sea otters are rarely seen outside waters more than 120 feet in depth, this is probably about as deep as this animal is capable of diving. Swimming speed is slow compared to that of a seal or sea lion, rarely more than eight miles an hour, and then only when badly frightened. The sea otter often swims on its back.

Sea Otter [*Enhydra lutris*]

A Life Story

PRELUDE

Since both the eared seals and the otters are probably descended from remote mustelid or weasel-like ancestors, the two kinds are related, but there are still radical differences between a sea otter and a seal. The seal, which is more completely a marine animal than the otter, probably left the land and branched off from its weasel-like ancestors some millions of years earlier than the sea otter, which is much more weasel-like. The main difference is that the sea otter has a tail which is used in swimming, while the seal has replaced any tail with flipperlike hindfeet. The sea otter also has very different molar teeth, much larger, wider and adapted for crushing seashells. To hold these teeth, the skull is much broader than that of the seal. The only seal with teeth like an otter is a very specialized one, the walrus, which has premolars adapted by their large size for crushing molluscs. The similarities between the sea otter and the river or land otter make it fairly certain that the sea otter descended by evolution from a land otter ancestor and began to become a sea mammal probably sometime in the Miocene Age, around twenty to thirty million years ago. However, it never became a complete sea mammal, able to brave the great ocean and its storms, but was forced by its small size and other limitations to stay in shallow coastal waters, usually taking refuge from storms on the shore or on rocky islands.

The Japanese hunted sea otters for many centuries, in their northern islands and in the Kurile Islands still farther north, and sold their beautiful warm skins to the Chinese at fantastic prices. The more primitive peoples of the Siberian and northwestern American coasts also killed sea otters for their furs and flesh for thousands of years. But the great modern exploitation of the sea otter for fur did not begin until 1742, when the remnants of the Russian-sponsored Vitus Bering expedition returned from the Bering Sea and the Commander Islands to Kamchatka in Siberia. The Danish commander Bering died on an isolated island now bearing his name, but his men brought back fabulous tales of the vast numbers of sea otters they had seen, and displayed their valuable skins. This news made the reckless, money-mad Russian *promychleniki* (fur hunters) rush to sail, often in poorly made ships, northward and eastward, hunting sea otters, first on the Commander Islands off Siberia, then up through the Aleutians to Alaska, and finally along the whole northwest coast of North America, wherever these shy and harmless creatures could be found.

The hunt for these skins was not only destructive of the sea otters, who were nearly wiped out as a species by the end of the nineteenth century, but almost equally destructive of the native populations and the cultures of the islands and mainlands where the fur hunters came. Thousands of the Aleutian Islanders were wiped out in bloodthirsty battles, and other thousands forced into semi-slavery as hunters by the Russians. So crushed by rape, murder and exploitation were the Aleuts that today a single full-blooded Aleut does not exist. Other tribes of the Alaska coast, such as the Kodiak Islanders shared the same fate. Only the powerful, numerous and warlike Tlingits of southeast Alaska were able to hold their own, even at one time destroying the town of Sitka, the ancient Russian capital of Alaska.

The Aleuts were especially useful to the Russians because the former had been sea-mammal hunters for many generations, and had developed an amazing proficiency with their walrus- or sealskin boats, or baidarkas, similar to the kayaks of the Eskimo. In these boats they could stay afloat even in rough seas, because waterproof skin flaps were tied around their bodies from the boats, and their skin clothing was waterproof too. When the extensive killing of the sea otters on the beaches drove these intelligent animals to seek refuge at sea, particularly in the kelp beds and among the rocky reefs, very dangerous to most boats, the Aleuts were able to pursue them in their baidarkas into these dangerous places.

Possibly only two things prevented the Russian promychleniki and their Aleut slaves, whom they took sea-otter hunting as far south as the Santa Barbara Islands, from wiping out the sea otters completely. One was that the Russians exploited their Aleuts with such cruelty, even driving them out to hunt during storms, that many of the best hunters were lost; the other was that few full-blooded Aleuts remained after two generations, because the Russians had mated with many Aleut women, and the half-breed and quarter-breed sons showed no stomach for the rugged and exhausting life of their fathers and grandfathers. Thus the last sea otters were able to retreat into difficult places among knife-edge reefs and rocks by the sea, where the new generations of hunters did not care to pursue them, and a small nucleus of otters remain alive, some in Alaskan waters, some on the California coast.

It is a tribute to the intelligence of the sea otters that they constantly changed their tactics and living places to escape their ruthless enemies, first retreating from the beaches to the kelp beds, and then to wild rocky reefs. At last in Alaska and on the then-wild and largely uninhabited central California coast a few found particularly unassailable rocky areas where they developed a coordinated system of signaling the approach of danger and remained almost completely hidden from man for several decades. Finally, in some remarkable way we cannot yet understand, their intelligence made them aware that most men had agreed to let them alone. It was in 1911 that four nations in the Pacific region—the United States,

Great Britain (including Canada), Russia and Japan—
agreed to an international law preventing further killing of
the endangered sea otters. It was in the late 1930s that the
sea otters, apparently feeling this cease-fire was really
holding, began to come out of their hiding places.

SEA SPIRIT

We can call her by the ancient Kwakiutl Indian name of
Quo-mogwa, meaning "Sea Spirit," and keep the first part
of the name only, to make it short. Quo was born in one
of the great kelp beds found off Point Lobos in California
near Monterey. Her mother, supported by the inflated
floats or bulbs of the kelp plants, twisted and writhed in
labor, moaning a little to herself, until the baby otter was
thrust by the muscular contractions from her loins. The
mother bit in two the natal cord and then turning over
on her back floated high on the kelp floats, drawing her
baby out of the water onto her belly and holding her
there with her two front paws to dry in the sun and summer breeze. Quo cried once piteously, then subsided as the
mother began to groom her, using both her tongue and the
sharp little claws of her front feet as combs to make the
fur take life and air. Unlike the dark black fur of the
mother, Quo's was yellowish brown and thick enough to
give protection against the fog-cooled breeze. After the
first grooming, the mother directed Quo's mouth to the
tiny nipples hidden in the fur of her breast, and the baby
began to draw in the rich milk that would bring her
strength and growth.

So far as we know only one baby is born at a time to
a sea-otter mother. If two were born at once one would be
almost certain to die, since the mother has room for only
one to be held and groomed on her belly, and this almost
hourly grooming is vital to the baby's survival.

By the second day Quo was strong enough to be left

alone by her mother, rocked into sleep by the waves moving in long sleepy rills through the kelp bed, and safe from man, beast and sea. The kelp bed was too thick for sharks or killer whales, too far out for much chance of a gunshot from the land, and the air bubbles in Quo's fur made her so buoyant she would have had difficulty sinking whatever happened.

Away from her baby the mother was down along the rocky and occasionally muddy or sandy bottom of the shallow coastal sea, hunting for sea urchins, clams and other shellfish. She tore these from the rocks with her claws, placed them in the furry pouch across her chest, then returned to the surface and lay on her back in the sea among the kelp plants, a little apart from her baby. Like a meticulous gourmet, she spread out her meal of shellfish and urchins two by two on her chest, hitting each set of two against each other to break the shells, then scraping the contents out with her curved front incisor teeth. When she found a clam or other seashell too hard to break, she took out a stone she had also brought up from the bottom in her pouch and used it as an anvil on which to pound the shell. Such use of a tool is very rare among animals.

Quo had started life as the smallest of all living sea mammals, for a baby sea otter weighs only about five pounds at birth; but in a few months she had grown to be twenty pounds and was learning to make her first shallow dives. Up to that time she was without doubt one of the most pampered babies in all the world of water, for her mother, aside from her occasional dives for food and her pauses for eating, was constantly with Quo, holding her on her belly and either giving her milk or grooming the fine and beautiful fur, a grooming that often lasted for more than an hour.

Quo at five months of age was a sprightly little girl otter, with a beautiful golden coat of fur that had replaced the more dingy yellowish brown of her birth. She was already making her first tentative attempts at play both with her mother and with other small otter cubs. During the first months the mother had been very jealous of her baby, growling at any other otter who came near her part of the kelp bed; but gradually contact with other mother otters was beginning to develop, the mothers forming a community in the kelp that brought their youngsters together.

Quo had little difficulty learning to swim, even shortly after birth. Her natural buoyancy made any danger of a prolonged ducking remote, but now she was getting anxious to follow her mother on her mysterious journeys under the sea. Her first attempts to dive were more comical than effective because of her very buoyancy. She would turn her body up to try to head downward in the water and find she was merely paddling air with her hindflippers. This frustrating experience was not overcome until she learned to use her front paws, with their curved retractile claws, to seize the underwater stems of the kelp plants and pull herself downward. But she had hardly learned to do this before she ran into another difficulty—how to hold her breath correctly. She was forced, sputtering and gasping, to the surface, spitting out water and kelp leaves! It was her play with other young otters that remedied this. When the youngsters chased each other about through the water and among the kelp plants in games of tag, hide-and-go-seek or king-of-the-castle, they were constantly ducking under the water and gradually they got used to holding a breath long enough.

Quo began to make fairly effective dives as she grew older and stronger, and she was fascinated by the underwater jungle of interlacing fronds and stems of the giant kelp. Among the beds was a maze of underwater channels and hidden passageways, the sunlight glinting down through the mass in long shining rays and making little rooms filled with light. Others appeared as dark caves of mystery, through which a sea otter pup was just small enough to find her way and travel on exciting journeys. Sudden encounters with other otter pups turned into games of chase-and-be-chased that could go on for hours, interrupted only by rushes to the water surface to get air. No better exercise place for young otters could possibly be imagined, since it was almost completely safe from all predators and designed to strengthen swimming and diving ability, length of breath, and agility. And there was always a fond mother near, watching, ready instantly to respond to a call for help or comfort when young Quo broke the surface. How marvelous, after a few hours of chase, mock combat and exploration, to rise to the surface in a state of near exhaustion and find a mother there waiting for you. Onto her stomach, as she floated on her back, Quo could crawl contentedly, find those remarkable nipples and start drinking the finest food in the world—rich and warm sea otter milk. Filled at last and absolutely content, she would be gently shifted off onto a thick bunch of kelp, where she could peacefully and blissfully go to sleep, rocked gently by the waves.

All through that first summer and fall of her life she found the sea moderately quiet, the waves singing on the rocky shore and hissing into the caves with only a mild roar. In November, however, the seas turned high with the first storm. But Quo had grown large enough by this time and was a strong enough swimmer to face the waves, though at first she was badly frightened and bewildered. Her mother, swimming beside her and touching her now and then with a gentle flipper, taught her how to ride up and up to the great green curling top, coast down the other side, holding her breath while the fierce wind blew spume into her mouth and eyes, then catch a deep breath in the slight calm at the bottom of the wave trough. She also learned how to swim a course across the wave front that prevented her being swept onto the rocky shore, where the seas were curling and crashing with roars that seemed to shake the world.

Modern sea otters on the California coast spend practically their whole lives in the shallow sea and among
the kelp beds, rarely going onto the land. This is probably
a defensive adaptation from the days of terror when the
only otters that survived were those who stayed away
from the land where hunters waited and watched with
guns ready. They even learned to move in such a way
among the kelp beds that they appeared like the kelp
leaves and floats.

The stinging spray, the wind whistling by their ears, and
the waves roaring and crashing on the rocks told Quo and
her mother when they were getting too near the land. The
two would then turn seaward in a burst of swimming,
able to surmount the worst storm because their hindflippers and tails were so perfectly adapted to the task of
driving them through the water. In the deeper waters a
mile from shore, they could dive down and down for
perhaps four or five minutes of rest in the depths, where
the pull of the storm made only a slight quiver in the
dark waters, and its sound a far-off murmur.

In the second half-year of Quo's life she learned to eat
solid food, for her mother began to give her bits of the
shellfish food that she brought up from the bottom of the
shallow sea near the coast. Quo particularly loved the
delicious great red abalone. Her mother, swimming thirty
feet or more below, would break open the shell with a
large stone, for the great seashell held to the rock with a
powerful suction. Even a powerful man with a crowbar
finds it quite hard to dislodge abalones from the rocks.
After Quo's mother had broken the abalone shell she
used her sharp claws and curved front incisor teeth to
scoop out pieces of flesh and put them in her chest
pouch to carry to the surface. Lesser shells she could
sometimes crush with her very strongly built and powerful molars.

By the following June, a year from the time when she
was born, Quo was herself diving thirty and forty feet for

food. Inexperience often resulted in bringing up inedible things, such as too-young sea urchins, too-small clams, and even pieces of rock, and her mother continued to feed her well into her second year. Sea otter females mate only about once in two years, and take care of their young for most of the interval. Quo was nearly two years old when her mother finally left her, one warm spring day, slipping away to seek a new mate so quietly that Quo did not notice she was gone until her frantic and vain searching finally convinced her. It was, however, not a complete separation, as the two females would meet again later, hunt together again, and include themselves in a social group, the so-called sea otter "raft." This is usually made up of either males or females, or females and their young, but rarely of the two adult sexes together.

After her mother left her, Quo had a difficult time for several months. Her clumsy attempts to find food were not efficient and a lot of the food she ate was of little value. One day, in desperation, she began nosing into the sand along the bottom, at a depth of about eighty feet, and scared up a sole, a pale whitish fish with a very flattened body that hides in the bottom sand. She seized it with one snap of her jaws, tore it to pieces and swallowed it ravenously. It was so delicious that she began to nose more frequently along the bottom to find similar fish. She found enough of them to turn the tide against starvation and began to enjoy a fuller stomach.

It was during such a hunt in her third summer that she became aware of a great dark shadow looming near, and turned sideward to see a twelve-foot-long hammerhead shark closing in on her. The creature was so strange-looking, with cold fierce eyes on the two hammer ends of its head, that at first she couldn't believe it was real. Then she caught sight of the great sharp teeth as the shark opened its cruel mouth, and she knew! She twisted to escape and headed desperately for a nearby kelp bed, but the shark was far faster, and sliced through the dim green water like a projectile. She made one more swift turn, as two razor teeth ripped lines of blood along her side, and the huge mouth yawned fiercely to chop her in two. At that instant something hissed through the water, something long and snakelike, but much straighter. The spear from an underwater spear gun in the hands of a scuba diver went straight into the shark's open mouth and through its roof into the brain. The shark writhed and twisted violently, but Quo was off, swimming desperately for her kelp bed. She was totally unaware that her life had been saved by a man, her species' most deadly enemy in the past, but now an unwitting friend.

Soon she became more expert in the other ways of hunting used by her kind. By watching the others she learned to pick up a handy-sized stone on the bottom of the shallow sea near the coast and use it to break open the shells of abalones, mussels and clams to get at the delicious contents. Some clams and abalones she was able to tear away from the rock surfaces with her sharp claws, and carry to the surface in her chest pouch with a suitable stone. There she would lie on her back luxuriously in the midst of a kelp bed, letting the swell of the ocean rock her gently, while she used her stone either as an anvil on which to pound her shells or as a hammer to break open the other shells spread out on her chest. Out of each piece of shell she would scoop the delicious meat with her front incisor teeth.

Another luxury was to float on her back in the midst of a raft of her friends, mainly females, with her tail and legs pointing straight up to the sunny sky. It was perhaps a way of sucking up sunlight and needed vitamins into her skin and blood, but it was obviously most enjoyable for she would do it for hours, chortling back and forth with little grunts and whines to her fellow females, in what might be called a mild form of gossip!

Gradually, then more rapidly, her time came to feel adult and interested in the opposite sex. All kinds of strange feelings churned inside her, but among them was also fear of what was happening to her. When two males discovered she was in heat and began fighting over her, she became very excited and watched them lashing about in the water, first one up, then the other, biting, growling and tearing. But when the smaller and younger male retreated, dazed at his defeat and streaming blood, and the older and larger male came eagerly toward her, she bit and snarled at him. Again and again he tried to seize her by the neck, a common practice of weasel-family males at the time of mating, and she would whirl away from him. At last he did get a grip with his teeth on the back of her neck, but he was too eager and rough and bit her deeply. She screamed with rage and pain and twisted violently out of his grip. Then she swam swiftly into a kelp bed, the male nose to tail behind her. Down underwater she dashed and up into the thickest part of the bed where she found what she wanted, a passageway between the heavy stalks of kelp so narrow that he could not follow her. Losing him, she swam far to the south, the wound in her neck throbbing and driving her on.

It is very strange, this need of the male sea otter to bite the female, often quite deeply. It is actually harmful to the species, as many a female dies from infection of the wound, especially if the coastal waters are at all polluted. Fortunately for Quo, the waters south of Point Lobos, because of the wildness of the coast, are comparatively clean, and her wound healed quickly. But she had been so frightened that she avoided males until the next time she was in heat, when, with some feminine wisdom, she selected a young male who, though ardent, was far gentler with his bites.

Quo played with him in and out of a kelp bed for three days, sometimes answering his embrace, sometimes dodging away and playing a game of tag that often frustrated the eager male. But at the end of the three days, she felt she had had enough male company, and watching him carefully until he dove for food deep underwater, made off by secret ways where he would not find her. Such are the brief romances of sea otters!

(*Note:* Sea otter males usually travel together in their own rafts until they find females in heat and are lured away by their scent. Otherwise they have nothing to do with the family life of their species, being free roamers who mate with any female at hand.)

ORDER

PINNIPEDIA

(Seals, Sea Lions, and Walruses)

1920593

This order is closely related to the order Carnivora, and probably derived from common ancestors with the family Mustelidae, or weasels. A very ancient fossil seal (Semantor macrurus of the lower Pliocene epoch in Siberia) had a long slim tail and seemed about halfway between a seal and a weasel. Though the skulls of both eared and earless seals approach the weasel type, especially badgers' and otters', in appearance, some scientists (particularly McLaren and King) think the eared seals are more likely descended from bearlike ancestors. Some have even suggested dogs as ancestors. In any case the ancestors of modern seals, sea lions and walruses apparently long ago lost their tails (unlike the modern sea otter whose tail flattened into a useful swimming oar and rudder) and developed their hind legs into flippers for propulsion through the water. Both the otter and the seal differ from the whale, who developed a nonbony horizontal tail called a fluke, which became its main source of propulsion, and who let its hind legs atrophy, as is shown by the remnants of these legs found in whale skeletons.

The Otariids, or eared seals (sea lions and fur seals), have records of ancestry going back around thirty million years to the Oligocene epoch, while the Phociids, or earless seals, are more specialized and more recently developed, with ancestors found in the fossil records of the Miocene epoch, about twenty million years ago.

The Pinnipedia developed fat, or blubber, under

their skins as a guard against cold, very much as the whales did, but unlike the whales they retained hair on the skin. The fur seals even developed a special underfur of great fineness, with air spaces to keep out both cold and water, similar to that of the sea otters. The blubber in Pinnipedia not only protects against cold, but provides buoyancy, padding and reserve energy, both for the female during the time she is giving milk, and for the male, especially the eared seal, who goes on long fasts while establishing a harem.

Pinnipedia obtain most of the water they need from the food they eat, which means their blood has less salt than sea water. Their urine, on the other hand, has more salt than sea water, and voiding may be how they get rid of excess salt taken in from the sea through the open mouth.

Pinnipedia generally have only one pup per female, as do whales and sea otters, it being very difficult for a sea mammal to protect and feed two young at once. By delayed implantation of the sperm in the ovary, some seals take ten, eleven or even twelve months between conception and birth, and can thus fit the birth in with the best season of the year for rearing the young. As with whales, the milk is extremely rich, around forty-two percent fat and designed to produce protective blubber in the young at a very rapid rate and strengthen them for fast swimming.

Whether all Pinnipedia use echo-location (pulsing sounds like sonar in submarines and bats) to find their prey in the sea is not yet known, but there is strong *evidence that many species do, as blind earless seals have been found to be fat and healthy, good evidence of their finding food without the need of eyes. And some sea lions have been recorded making sounds like echolocation pulsations. All appear to be animals of comparatively high intelligence, though without the sensitivity that causes moodiness and even apparent suicide in some whales.*

FAMILY

OTARIIDAE:

Eared Seals, Including Fur Seals and Sea Lions

This family is distinguished among seals, or pinnipeds, by having small but visible ears and the ability to draw their hindflippers up under the body or beside it in such a way that they can be used for waddling or even doing a queer swaying gallop for travel over land. They can also use the flippers for climbing proficiently up over rocks, something the Phocidae (hair, or earless, seals) find impossible to do. Another peculiarity of the Otariids is that the first and second incisor teeth are small, but the third is long and shaped like a canine for use in fighting and for seizing food.

Northern Fur Seal [*Callorhinus ursinus*]

Northern Fur Seal

[Callorhinus ursinus]

Description   Male up to eight feet in length, weight up to 750 pounds, generally around 600. Female up to five and one half feet long; weight up to 150 pounds, generally around 100. Male dark brown; female and juveniles more grayish, sometimes silvery or golden; newborn pups are mostly all black. Note the short nose of the male and his markedly steep forehead (sometimes bulging) for a seal, equalled only by the male California sea lion. Hindflippers of both sexes can be brought forward under the body or beside it for climbing on rocks. Ears are easily seen. The fur of the fur seal is especially distinctive, since beneath the coarse guard hairs is an extraordinary warm and thick coat of very fine fur, impermeable by either cold or water. On hot days on land the seal keeps cool by fanning its large black naked flippers, while in the water on cold days it may lie on its back in the sea and hold the flippers out of the water to prevent heat loss.

Range and habitats   The Bering Sea is the northern range of the Northern fur seal, with the breeding grounds on the Pribilof Islands in that sea its main center of summer activity. During the fall it may migrate southward along the Pacific Coast of North America as far as Baja California, spending most of the fall, winter and spring (when it migrates back to the Aleutian sea area) as a pelagic, or open-sea wandering, animal.

Food and locomotion   While very young the fur seals feed mainly on molluscs and other rocky shore life, but the adults are primarily hunters of fish and squid in the open sea. Most of the fish they eat, however, are not of the commercial varieties. Speed in the water is up to about fifteen miles per hour, with occasional faster bursts. Most swimming is done by both the twisting of the body and the paddle or winglike stroking motion of the foreflippers, the hindflippers being mainly used for steering. Strangely, high speed seems to be done by body-twisting, working with the back flippers.

A Life Story

PRELUDE

For millions upon millions of years the fur seals of our world, which include several different species, have had hidden refuges, islands so far from land and sheltered by the immensity of the great seas that no land carnivores could reach them. Some were and still are found in the South Pacific, off the coasts of Australia and New Zealand, some on or off the desert coasts of southwest Africa, or on the islands off the west coast of Mexico. But the most important of all, the supreme heaven and paradise of the fur seals, has for ages been the Pribilof Islands in the Bering Sea about four hundred miles from the mainland of Alaska. Each year as many as two to three million fur seals have repaired to these islands in summertime to raise their young on land, free from the fear of killer whales, sharks and polar bears, free, until a mere two hundred years ago, from man, the most terrible carnivore of all.

In 1783 the Russians under Pribilof discovered the islands named for him and immediately let loose a wanton killing of this animal for its valuable fur. The seals might have been wiped out within a century if the Russians themselves had not come to their senses and set a limit on killing, in order to preserve enough seals to maintain a steady yearly harvest of their furs. When the Americans took over Alaska in 1867, they in turn were so anxious to reap a fortune from the furs that they ignored what had been learned, and again started a wholesale killing. By the early nineteenth century the population of Northern fur seals had dropped below the hundred thousand mark, and the species seemed doomed. Only the strenuous efforts of conservationists, who recognized the danger to a magnificent species of mammal, finally caused the U.S. government and later Japan, Russia and Great Britain (including Canada) to restrict severely the killing of fur seals so as to regain some of their former numbers.

In the old days the sealing parties used to rush up the beaches of these islands waving clubs and shouting ferociously while they drove the seals inland to where they could cut them off from the sea and club them to death. Most of them delighted in this killing, but a few, noting that the mothers often threw themselves between the man with his club and their young, and seemed to cry for mercy, became sickened of the slaughter and gave it up. Today only selected bachelor seals, all in the three-to-five-year age group, are harvested each year in the Pribilofs to keep the fur seal industry going. Perhaps one day we will recognize the fur seals as the intelligent and feeling beings they are, and see greater recreational value in a national park to observe them than in the sale of their furs. Then the killing will stop.

HARL THE KING

It was late June, after nearly a year of absence, that the beautiful silvery fur-seal female saw the beaches of Lukanin. She was heavy with a pup as she approached St. Paul Island, one of the Pribilofs, swimming sluggishly after her long journey out of the south. When the remembered roar from the beaches reached her ears as a distant rumbling, she twisted suddenly forward through the sea with renewed vigor. As the roaring and muttering grew louder, the familiar smells began to reach her on the sea breeze: the smell of live seals sweating in the rare sunlight of a beautiful day, the smell of excrement and the fetid smell of death. None of these smells troubled her, nor did her memories of the packed bodies on the beaches and sharp nips given her by angry mothers when she passed too near their pups. The great pageantry of the place with its coquettish cow seals and pompous and fierce bulls, its anxious young adolescent seals trying to get through the harems and rookeries to their own special meeting and sleeping places, its jet-black babies shrilly calling for their mothers, its mothers as anxiously calling their children, all the confusion, the battles and retreats of harem masters and their challengers, even its wounded, dying and dead, all were part of a great mass rite of the fur-seal people, and thrilled her heart with the joy and anticipation of joining it. Here her kind were gathered from the four corners of the Pacific in this place of new life, the start of a new generation each year, a place to dance together in the waves, to hear the roaring on the rocks echoing the roars of the great bulls, to feel the deep mists of morning on the wet sands, to feel this was home, the place where she was born.

As she came in toward the beach, skimming over the waves, her eager eyes searched the sands and rocks, seeking, as every female fur seal does who comes back to the Pribilofs in summer, a bull who would be her lord and master. The vast difference in size between the male and female fur seal is common among mammals of the species that have harems dominated by large males. This may have made her seek a bull seal with a thick neck, the sign of great muscle power, of one who would dominate and outbluff other bulls and eliminate a lot of fighting. Who knows how each female fur seal makes her choice, since every harem master on the beach stakes out a territory in which he has at least one female and often more than twenty. Maybe some females choose the more lonely bulls on the outer fringes because they know their harems will be less crowded; and maybe some choose a particular bull because he has taken charge of a very smooth bit of sandy beach. Perhaps most of all many choose a particular male because in some odd feminine seal way, not entirely unlike that of women, each simply likes his looks and feels enamored of him!

The silver seal saw a great bull, brown in color like the one she had chosen the year before. She noted he was not very scarred by battle, a sure sign that he was so huge and powerful-looking that few other males had dared to challenge him. Further, she liked the quarter acre of beach he owned for its nearness to the sea and smooth sand, and feeling tired from her long journey and ready for a rest, she heaved herself out onto the sand with a sigh of content. The bull noted her coming, lifted up his body and puffed out his muscular neck to make it look as big as possible, at the same time giving the whickering call that told her she was his. Majestically, then, he moved down toward her, rocking back and forth as he came, urging her first with grunt and then with nose to come further up on the beach where he could protect her from other bulls. Perhaps he was telling her in his own way that he had come specially early to seize this fine beach for her and the thirty-five other seals of his harem, that he had been fasting as all the harem bulls do for more than a month, and that he would fast for many more days to come. If so, she paid no attention to his claims. She had made her choice, and now all she wanted was to rest, and, tucking her head to one side, she fell soundly asleep.

Her resting did not last for much more than eight hours in the almost constant light that surrounds the Pribilofs in midsummer, when the sun dips below the horizon for less than an hour. Soon after the gleaming midnight, she began to feel the contractions and pains that warned of a new life coming. It became her own private struggle during the next two hours, her solitary meeting with destiny, for no one else paid any attention to her; not the bull, looking for new conquests and possible rivals, nor any of the other female seals, each intent on her own pup or her own birth pangs. It is a cruel life among the fur-seal people, for they have not reached the stage of cooperation found among an older sea folk, the whales.

At last she gasped, moaned and writhed, as a new body burst from her own and lay beside her on the dark sands, wet and shining, still attached by its umbilical cord. The newborn pup was a very unusual specimen. Unlike the common black-colored fur seal, he was beautifully marked with a blend and contrast of black and white, an effect so startling that, though it would attract attention in the sea, it would also later excite fear and caution. Its appearance was different from anything else likely to be seen among the waves, except for the mighty black and white pattern of a killer whale. Only about once among a hundred thousand births of fur seals is such a creature born, called piebald, and probably as rare as a pure white seal. Harlequin, as we shall call him, or Harl for short, had the bloodline of one of the mightiest of all fur-seal bulls and generations who had fought successfully the dangers of land and sea. But his chance of survival beyond the first year was perhaps one in five, for terrifying are the hazards that face all the newborn seals on the Pribilofs, as we shall see.

As he sucked the rich (thirty percent protein) milk of his mother, he was blissfully unaware of coming dangers. The milk she was giving him, and which poured out whenever she was with him in the next three to four months, was fortified with both the growth-producing power to make him increase rapidly in size and the resistance needed to avoid the diseases and parasites that might beset him. In the fetid unsanitary conditions of the fur-seal rookery, many pups fall ill from some disease, particularly from the attack of hookworms, which abound in the droppings. It is strange that some pups and other young seals were apparently able to pass out these worms quickly before they did much harm. Throughout the rookery were hundreds of pitiful dead bodies of pups who had not made it. These bodies are cleaned up almost daily either by the island foxes, who along with ravens and some sea birds act as scavengers, or by the biologists of the island of St. Paul, in an effort both to keep the beaches cleaner and to determine statistically the percentage of each disease or parasite that causes death.

Harl had other built-in inheritances and instincts from generations of seagoing ancestors that would serve him in many an emergency to come. At the end of his first week of life, surfeited with milk and sound asleep on the beach, he lay in the path of the great bull fur seal as that ponderous animal moved in lordly manner through his harem searching for females in heat. Totally oblivious to young seal pups, intent only on his mission, that grand patriarch would have crushed Harl like the egg of a black sea tern had not some sense warned the youngster at the last possible second to roll down the beach and out of the way of the approaching steamroller.

The bull passed on, and finding the nose of Harl's mother redolent with the mating smell, snorted in eager anticipation as he began the mating dance round and round her, while she snarled and nipped at him in preliminary protest. His eagerness finally aroused her, however, and soon she was giving the final mating signal of spreading the brilliant pink of her rear parts, and he almost crushed her body with his seven hundred pounds, and made her eyes bulge in the culmination of the mating act.

Soon after this she left her pup alone for the first time, called by hunger and the need to build more milk. Eagerly down to the waves she went, washing the mud and excrement of the beach from her lithe and silvery body, as she flashed down to the depths, where the fish fled before her in dazzling streaks of light.

Harl, finding himself suddenly alone when he raised his head from a deep sleep, at first cried piteously for a mother who never answered. His grief subsided to a soft whimpering until he found another equally deserted pup, and the two foster-mothered each other, snuggling together and sucking at each other's whiskers in temporary imagined suckling in lieu of something better. After four long weary days, for the sun barely sets at all in early July among the Pribilofs, Harl heard a cry that made him quiver and then answer with all the power of his small lungs. And she came, she came, the bringer of all things good!

Again came the blissful days of lying beside her, sucking in that all-providing milk between wonderful long naps, and growing meantime bigger. Then followed another long absence as his mother set up the accustomed rhythm of sea and land. During this second absence a great golden bull, roaring his challenge and puffing forward for combat, came out of the sea and over the rocks, seeking the huge brown harem master. The two colossal creatures at first circled each other slowly and warily, getting closer, then standing a moment side by side with heads cocked and teeth bared in the ancient fur-seal ritual in preparation for battle. Then the golden one lunged like a streak of light for a vulnerable flipper. But the owner of the flipper was a veteran of a hundred or more grim battles, and he turned as a ballet dancer turns, his own head flashing out to seize the golden one's thick neck. They were nearly equal in size and the snakelike strikes, the tearing of skin and flesh, the dripping of blood from deep gashes, the rising and falling roars, the panting pauses to gather strength, the pushing and hauling back and forth over the rocks when each had a deep grip on the other— all this lasted for more than half an hour. But it was the harem master's territory and he was hanging on and fighting as if he could go on forever. At last the golden one's nerves snapped as a cable does when jerked too strongly by the drop of a hundred tons, and suddenly he turned in a lightning move to flee over sand and rocks down to his escape, the sea, the harem master galloping behind him. But the anger of the defeated bull lashed out as he ran, at two little bodies huddled together for comfort. One lay paralyzed with fear before the bull's rush. The other moved at the right instant, rolling sideways. One was thrown high in the air by a nip and then a toss from that powerful neck. A thin cry gurgled from the baby's mouth, and was silenced as the small body smashed to crumpled lifelessness on a rock. Harl, safe to the side as the great bull dashed furiously into the sea, lay still for a while before he began to cry for his mother.

As Harl grew larger he learned what all pups must learn to stay alive in the rookeries, to move to the shelter of large rocks where the bulls did not often pass, and eventually to move up on a higher part of the beach where pups lived separate from the harems, found and fed by their mothers hauling in from the sea.

He had known instinctively how to swim from the day he was born, but his head was so much bigger than his body at that time that he would have capsized in deep water. When the time came to follow some older pups down to the edge of the sea, he found a high tide pool about a foot deep in which to splash and play and make swimming motions. Another day saw him in a bigger, deeper pool, sliding down a rock into it and sloshing and half-swimming in the water. Growing more confident with each new victory, he was soon diving through the first breakers and feeling that ultimate thrill of the young seal, the swish of water about him made by his own flippers. Now he was ready for that most ancient and exciting of all seal pups' games, king-of-the-castle. Up he would climb from the water onto a rock, partially washed by the waves, and toward him other young seals would swim, chirping their challenge. Surrounded, he would try to cock his head with the proud look of a great bull, then push and shove them away with flippers working furiously and his head driving like a battering ram to knock them down. At last, in a frantic swirl of bodies, he would be dragged unceremoniously from the rock by a bigger pup, who would be the next king of the castle. So muscles were strengthened daily, while the bodies were built by the rich flow of milk.

Gradually the milk was supplanted by fish from the sea, regurgitated from his mother's mouth, in the closing month of the seal island summer, giving him the taste for fish he needed when he would have to hunt them himself through the far waters.

In a last blaze of glory, when the sea turned with the lengthening night into flames of fire from the millions of tiny luminescent animals and plants that swam in it, he danced with other young seals in the waves, the bright light streaming and flashing about his young supple body as he breached the wavetops and leaped high in the air, flashing up and over like a dolphin. How glorious the feel of his muscles in action amid the deep songs of the sea, how wonderful the streaming water, and all about him his companions, hundreds and hundreds of them, leaping and playing in unison!

At last, as the cold winds began to come down from the Arctic in late September, the glorious days and nights were over and the terrible time came as it comes to all seal pups at the end of their first summer, when mother no longer comes up out of the sea with her delicious milk and her regurgitated fish food, when all his calls are unanswered, and he sees the older seals hauling off the rocks and down into the sea, heading southward in the great migration of their kind. Dancing in the waves and sometimes braving the deeper water, he has already had the occasional joy of seeing a fish swimming before him and has dashed forward to catch it, or found other edible life, such as shellfish and shrimps, in the tide pools; but now he is hungry and there is no one to feed him except himself. He sees a group of his comrades going down the beach into the waters, their purposeful movements showing they have, at least unconsciously, faced the moment of truth and joined the deep-sea-goers. Into the sea he leaps after them, fearful and wondering, for life's greatest adventure for a young seal is upon him.

Instinct tells him to swim south, but whether he moves only by the sun that has begun to dip in early November low toward the southern horizon, or also follows the point lines of the constellations at night, we may never know. He weighs about thirty-six pounds when he starts, but he will be lighter and far wiser by the time he reaches the British Columbia coast, two thousand long sea miles or more away. Instinct tells him how to sleep, his fur filled with air bubbles that keep him buoyantly riding the wave-tops, resting on his back, his nostrils coming up automatically out of the water to breathe air at the proper intervals, his sensitive whiskers alive to movements in the water around him, his ears listening for sounds and his brain sorting out their meanings.

He is fortunate to find a group of others travelling south, another pup or two, some yearlings, a two-year-old "holi-schuckee" or bachelor. Most fortunate of all, there are two older females, for it is from their reactions to sounds and other sensations in the water that he will learn sea wisdom. It is from them that he will learn the peculiar high clicking sound used by the fur seals to echo-locate schools of fish and follow them down and catch them in the darkness of night when sight is of little value. Several kinds of animals have developed independently the same use of sonar that guides bats in their nightly pursuit of insects. Since the fur seals usually travel or rest by day when on migration, and hunt their fish food by night, this sonar is vital to them. But Harl was completely new to this kind of hunting, and he had to make many frustrating misses by sonar before he began to get the feeling of how to use it.

They came to Unimak Pass in the Aleutians, the straits through which many thousands of generations of seals have begun their yearly migration south across the Pacific, and now he suddenly found himself caught in a North Pacific gale. The waves rushed toward and over him two and three stories high, their tops racing southeastward with the spindrift white on their crowns, and the wind fiercely whipped the spray like needles against his face when he came up to breathe. It is a time of terror and helplessness for a young seal, with no time to hunt food, only to fight desperately to stay alive, and many die in the rough waves. But Harl watched the older seals and learned to ride up and up to the crest with his head and nostrils turned downwind, and to lie and rest for a few moments in the troughs before the next wave could catch him.

Two hundred miles they rode the storm eastward, until rising on the top of a great curling sea they viewed the Shumagin Islands and gradually worked their way into the lea of the land where a bay could shelter them from the storm. What a relief to lie in the quiet waters and sleep for a while, then rouse out of sleep for a night of fishing. When the storm passed and they were out in the Pacific, once more migrating south, they heard the distant talk of killer whales—a clicking and shrill chirping that ominously told the seals that the whales were echo-locating for other sea mammals to kill and eat. These chilling sounds warned the wise old females to turn swiftly to reach shallow waters where the killers would be afraid of the rocks. The sounds came louder and louder, however, and before they reached the rocks the great black-and-white high-finned sea mammals were upon them.

The older female seals, knowing the ways of the killers, dodged quickly each time one of the ten-ton monsters struck at them with gaping mouth and long sharp teeth, the seals' bodies turning more sharply than those of the twenty- to twenty-five-foot-long whales. Harl saw a great black-and-white body rushing toward him at a speed far faster than his own, and fled in hopeless terror. At the last possible moment the monster beast swerved and seized another seal pup in its many-toothed jaws, swallowing him whole without a sound. Possibly he had seen the piebald pattern of Harl's body and thought this creature might be one of his own black-and-white kind.

A two-year-old seal was tossed twenty feet in the air by a killer bull, and almost cut in two as he fell screaming from the sky. Then, almost as quickly as they had come, the killers were gone and there was only blood in the sea, and a lonely seal pup looking desperately for those who remained alive among his fellows.

It was three days before he found them, a few miles off the Trinity Islands, south of Kodiak, Alaska, and it was the Alaska current, sweeping eastward and then south past those islands, that carried him and two other pups and three yearlings to find shelter at last from winter storms in the Inside Passage of the southeast Alaska coast. Here, among delightfully forested islands, the winds blew more gently, and he moved leisurely south till he had reached the British Columbia coast and the first snows of winter. In early December he had arrived at the lowest weight of his journey, twenty-five pounds, a tiny mite in the midst of the sea, but was learning fast how to make the right turns in hunting the small fish and squid of these waters, so that from now on his weight would go up.

Harl was beginning to pick up the rhythm of the fur seal, which is to go south in migration to less cold and fearsome waters but return in summer to the Pribilofs in the Bering Sea, first as a yearling and then as a two-year-old bachelor, or holischuckee, always growing bigger till he neared the final size of the great bull fur seal. But the rhythm varied in detail from year to year. That first year he lingered up and down the Inside Passage, learning many of its numerous channels and byways, its good fishing places, its island walls for protection against the terrible Pacific winter. In his second year he wandered as far south as San Francisco Bay, and in his third winter he adventured still more to the south, to the central coast of Baja California, in Mexico, which is about as far south as the Northern fur seal goes. There he met his first longer-nosed and smaller cousin, the Southern fur seal, a creature almost wiped out by man, but making a gradual comeback on some of the southern islands off Mexico and California. Since the two kinds could not understand each other, only stare curiously back and forth as they crested waves in the sea, they soon parted company.

Not long after this Harl, a brash and fast-swimming three-year-old, lost all brashness when danger came close and terrible. A giant white body pushed against him, even as it struck with deadly silent aim at another of his kind, a four-year-old female, too terrified to dodge in time. He saw the great white shark, all twenty feet of it, moving with twice the speed of the fastest seal, opening a maw lined with teeth almost half a foot in length, and closing them over the whole body of the fleeing seal. Another seal was bitten in two as it tried to wheel away, the shark swirling itself like liquid light, and swallowing the two pieces in as many mouthfuls. Harl saw the huge deadly monster moving through clouds of blood and fled like a frightened halibut, unaware that his own life had been saved for a second time by his peculiar color combination which made so many sea creatures, even his own kind, approach him warily.

Harl spent his fourth summer at the Pribilof Islands. The bachelors, from ages about two to seven, had their own beaches and places inland, where they lay and dozed in the sun for hours, played such games as king-of-the-castle or tag, and also practiced at fighting. They watched the great bulls at their ritual challenges and real fights, and copied their moves. It was during this play fighting, that sometimes degenerated into something close to the real thing, that Harl one day learned the trick of changing colors. His left and right sides had entirely different patterns of black and white, and he discovered that by suddenly swinging his body in a half-circle to present to an opponent the opposite side from that first seen, he could so startle the other animal that he got a chance either to knock him down with a sudden blow or seize a vital part of his body before the opponent could prepare himself.

As Harl grew older and nearer to bull size, he began to stay further north in winter, first in the Alexander Archipelago near Juneau, then north and west in the Kodiak and Shumagin Islands. Though it was colder there than he was used to, his thick layer of blubber and his now very thick and fine pelt of fur gave him all the protection he needed against the winter cold; and the fishing in these far northern waters was better than good.

At nine years of age, hard as nails and weighing six hundred and fifty pounds, he came up out of the water at Lukanin Beach on the Pribilofs, his eyes red with the rage of combat, and headed for a great black bull who had a good harem spot on the beach. Ordinarily a first combat for a young bull with a veteran of this sort would have quickly ended in death or near death for the youngster, but Harl had learned his tricks well, and he had learned to be a winner among the bachelors. He had also waited cannily till early July, when he was well-fed and boiling with energy, fresh from two months of leisurely fishing. The bull he faced had already been two months on the beach without food and water and had gone through other fights that had left their scar marks on his body and fatigue on his mind.

The two came up to each other side by side, muttering and roaring threats, cocked their heads and appeared to be considering what to do. As the great black bull suddenly lunged at him, Harl whirled his body so the other side came into view, making the bigger seal think he was suddenly facing a new opponent from a different direction. And in that moment of startled incomprehension, Harl struck for one of his enemy's front flippers and seized it, which the veteran would never have allowed had he been fully alert. Tearing at the flipper, Harl felt himself seized by the shoulder in a powerful grip that nearly lifted his whole body from the rocks, but the blubber was so thick there that he hardly felt it. Meantime the black bull was suffering terrible pain and the fear of losing a flipper, which could lame him for life. Suddenly the black bull tore himself away, leaving a part of his flipper behind, and rushed down to the sea bellowing insanely. The beach harem of some thirty cows had a new master, one who would be harder to beat than any other bull on the beach! The cows stared in astonishment at his strange color pattern, but they were not alarmed. On the beaches of Lukanin as well as on all the other fur-seal beaches of the Pribilofs, he who wins the battles wins the most mates and keeps them. It was the beginning of a long and successful reign for King Harl, as the sealers and the biologists came to call him, and the fighting bloodline of a new and powerful beach master would be passed on to many a fur-seal pup.

Southern, or Guadalupe, Fur Seal

[Arctocephalus townsendi]

Description Length of male up to six feet and weight up to 500 pounds; female around four feet long and weight up to 100 pounds or more. Both are covered with thick brown underfur of very fine quality, their front areas

Southern, or Guadalupe, Fur Seal [*Arctocephalus townsendi*]

marked with silvery guard hairs, which give a grizzled appearance when dry. The nose is much longer than that of the Northern fur seal, and the male has a sloping forehead, not with a prominent crest as in the Northern fur seal. The distinctively long and narrow hindflippers are characteristic. Voice is quite different from that of the Northern fur seal. Young seals have three unique calls: a high-pitched roaring, a sharp bark, and a guttural cough; this last is used in threatening other seals. Adult males have a deeper bark and roar; also a rasping and harsh "puff" sound to indicate the boundary of a territory. Females with pups give a long-drawn-out bawling cry, evidently for calling their pups.

Range, habitats and some habits   Southern fur seals were originally found populating the beaches of all the Santa Barbara Islands off the California coast and south to central Baja California. The present range is from Guadalupe Island off northern Baja California to the Santa Barbara Islands, particularly San Nicholas Island. The species was so badly decimated by the seal hunters of the nineteenth century that only a handful remained at the beginning of this century. These escaped persecution mainly by hiding in deep sea caves, whereas the former population had congregated on open sandy beaches. Somewhere over 600 of these seals exist today, but they still haul out and breed mainly in very rocky areas where there are plenty of caves to hide in. The result is that there is far less direct competition between bulls, who form harems in May to July on Guadalupe Island, and very little fighting. But young pups have a great time playing in the tide pools. Some of these seals remain throughout the year on or near Guadalupe Island, while others spend most of their time far out at sea, except during the breeding season.

Locomotion and food   Their swimming resembles that of other otarine seals, with the foreflippers used like wings or oars in the water; the speed is probably the same as that of Northern fur seals—up to fifteen miles per hour. Food is generally various fish and squid, caught in flight by quick dashes under the surface, but the young seals may feed on crustaceans and molluscs caught in the tide pools.

Otherwise the life of this species is very similar to that of the Northern fur seal described in the section before this one.

Northern, or Steller, Sea Lion

[Eumetopias jubata]

Description   Adult males nine to thirteen feet long, weighing up to 2,200 pounds or more; adult females six to nine feet long, weighing up to 800 pounds or more. Color mainly pale yellowish or yellowish brown, but varying from almost creamy to dark yellowish brown; flippers blackish brown; pups are generally dark brown, looking blackish when wet. The head of the male slopes back from nose instead of

Northern, or Stellar, Sea Lion [*Eumetopias jubata*]

forming a prominent bulge or crest as with the California sea lion bull.

Range and habitat   Found from the Santa Barbara Channel Islands north in winter to the Pribilof Islands in the Bering Sea, and in summer still farther north to St. Lawrence Island at the north end of the Bering Sea, but staying always south of the fringe of pack ice. Breeding is rare in the Channel Islands near Santa Barbara, but more common farther north at different remote islets and hidden rocky beaches along the coast north to Alaska, with the largest number probably on Año Nuevo Island near Santa Cruz, where more than 2,000 of these animals may haul out at one time during the breeding season. When away from land they tend to stay fairly near the coast and coastal islands, often hauling out to sleep, though they can sleep in the open ocean floating on their backs. Rarely go up rivers or into bays.

Habits   Ordinary swimming is by sculling with the large foreflippers, but fast swimming is done with hindflippers acting as propellers and the whole body twisting through the water. Top swimming speed is probably near fifteen miles an hour, and the body can often be thrown completely out of the water when porpoising. Favorite foods are squid and octopus and various kinds of fish. Shellfish are third, and various kinds of shrimps and crabs are probably fourth. Fast-swimming prey are caught by rapid dives and swift turns; shellfish are broken open by the powerful teeth.

A Life Story

PRELUDE

This largest of all the eared seals, or Otariids, is a great and magnificent animal, capable of diving beneath the sea for more than half an hour and to depths of 600 feet or more. It has not been studied as much as its smaller related species, the California sea lion, perhaps because it is so strong and powerful that it is not as easy to train to do tricks in a sea aquarium or circus, but it gives every evidence of remarkable intelligence and has successfully so far weathered the destructive impact of our civilization on the wild life of the Pacific Coast. This is partly because it hauls out or goes ashore mainly on rocky islets and isolated beaches surrounded by towering cliffs, places difficult for man to approach. It was, however, hunted extensively for its hide and blubber in the nineteenth century, and wiped out from some of its old hauling-out grounds and breeding places, such as the Farallon Islands, by the sealers. In this century it is returning to several of these places.

Like the fur seals, Northern sea lion bulls are very much larger than the cows, as is found among most animals that have harems. Though the harems of sea lions are not so tightly held together by dominant bulls as the harems of fur seals, the sea lions give evidence of a more lasting and cooperative form of social life. This is shown by the far greater care and solicitude shown by the sea lion mother for her cub than by the fur-seal mother for hers. The sea lion cub is usually nursed for seven months or even longer; indeed, sometimes a mother may be seen nursing two youngsters, one a newborn cub and the other a yearling. This enables the mother to teach her cub much more about successful hunting and the avoiding of enemies, and allows for greater social defense actions among animals that know each other so intimately. When I saw sea lions attacked by killer whales off the Farallon Islands, lookouts gave the cry of alarm as soon as the whales were spotted, and many saved their lives by getting ashore before the killers could seize them.

Northern sea lions were once much more common on the islands off Santa Barbara, but now they are rare, apparently because the water in that area has become warmer, and these large northern-oriented sea lions may dislike both the temperature and the kind of food animals now found in that area.

Northern sea lions are immediately distinguished from the California sea lions by their much larger size, lighter colors and the roaring of the males, which contrasts with the shriller barking of the California sea lion bulls. But the two species seem to get along well together when they meet on the beaches, partly no doubt because the smaller and more slender California sea lions can get quickly out of the way of their larger cousins, if there is any challenge, and so avoid unequal combat.

One of the most spectacular places to watch the Northern sea lions is in the Oregon Sea Lion Caves on the central Oregon coast, where an elevator takes you down inside the cliffs to observe them through natural openings. I greatly preferred walking down the winding trail to the caves in the old days, observing the bright colors of summer flowers and watching the seabirds spiral and dip over the great cliffs, but most people seem to want more convenience and comfort these days! In the caves you can see from close at hand the sea lions lolling and sleeping on the rocks, diving gracefully into the wave channels, the younger ones playing tag or king-of-the-castle, many scratching industriously for fleas and other pests, and the great bulls occasionally deigning to rear up and let out a thundering roar at one another.

Today there is a strong debate between fishermen and scientists as to whether the Northern and California sea lions are destructive to Pacific Coast fish, and so limit the catch of the fishermen. Most scientists maintain that the majority of the fish caught by the sea lions are the non-commercial kinds. However this debate may end, all sea lions have a tremendous recreational and educational value to the public in that they are among the few large wild animals that can be watched from close at hand in their natural habitat without disturbing the rhythm of their lives, so that their interesting habits on land and in the nearby sea can be studied. In some of the large modern sea aquariums they can also be observed swimming under water so gracefully and swiftly that it is hard to believe!

DAVID THE CONQUEROR

Perhaps many a human David wishes he had a Goliath to conquer and so prove his courage to some fair maiden, but the David in this story was a sea lion and the Goliath he had to conquer was no paper tiger, but a real-life Northern sea lion bull, well over a ton in weight and as nasty a tyrant of the beaches as ever ruled a rocky harem! And our David in this story had no sling or any other kind of man-made weapon to conquer his giant, but only his sharp teeth and his quick and supple body—perhaps also a supple mind.

David was born on one of the rock-shelved beaches of Año Nuevo Island, not far off the coast of Santa Cruz, California, in an old Northern sea lion rookery that may have been used by the sea lion folk for as far back as ten million years. From the long experience and inheritance of that vast length of time, his species had somehow developed extreme suppleness of body and especially neck, for catching fish in the sea and for avoiding danger from such monsters as sharks and killer whales. They also have powerful canine teeth, especially the bulls, for killing food animals, fighting enemies and exerting the dominance of the males on the beaches. A period of long-time nursing and educating of the young, which trains the little ones for life and the flowering of their intelligence, has been evolved by the mothers. The sea lions have a social system that gives protection against most enemies on land and some at sea, a high ability to communicate with each other by many sounds, smells and movements, very possibly a sonar system of echo-location that is very useful in catching fish at night or in deep waters, a protective undercoat of thick blubber that allows these seals to inhabit the icy waters of the north, and an ability to dive very deeply after food, especially octopus and squid.

David was born in early June amid the roaring of bulls, the barking of alarmed or angry cows, the yelping of Northern sea lion pups all about him, and the shrill screaming of gulls and other seabirds overhead. These noises and the animals and birds that made them, as well as the sighing, the gurgling and roaring of the sea, would soon seem almost as much a part of his life as the air he breathed, but at first he was confused and frightened. He was hungry and bleated for he knew not what, until his instinctively searching mouth found the teats of his mother and the wonderful warm rich sea lion milk began to stream down his eager gullet as he sucked. He had dark brown hair, was about two feet long and weighed around forty pounds, but he already knew how to pull his hindflippers up beside his body and use them as levers, almost legs, to waddle and scramble over the rocks and sand of the beach.

His mother was about one-fourth the size of the great bull who ruled the harem, and it was fortunate that she was an alert and eager mother, for twice in the first three days of David's life, she had to use her nose to push him quickly out of the way of the bull as he crashed and lunged his way over the rocky shelves, roaring and thundering a challenge to any prospective challenger, and as unaware of pups in his path as a bulldozer operator is of a mouse under his scraper! Most prospective challengers turned away and fled, with only a duck of their heads and a few parting roars; but an occasional bull of comparative stature scrambled up out of the waves onto one of the rock shelves and faced the harem bull with savage determination.

The hair would stand high on their necks as each raised his head and pointed his nose at the sky, roaring ferociously. Then the two suddenly would lunge at each other like striking rattlesnakes, their snarls and roars smashing across the beaches while many turned to watch (though most simply kept on about their business). It was not a battle quickly ended, as among the elephant seals, but a prolonged fight in which each tried to shake the other as a dog shakes a rat, and each strove to seize the other by a flipper, for a flipper hold can be completely crippling. But they were clever about keeping flippers out of the way, and as they lunged and bit and shook each other, blood ran down their necks and painted their bodies scarlet, and deep wounds opened like gaping holes. At last the harem master took such a hold that he could twist and throw the other bull down onto the rocks below the shelf where they were battling. The other bull groaned as he struck the rocky shelf and slid sluggishly into his Mother Sea, the waves washing off the running blood as he slowly and painfully swam away.

The call of the sea, the roaring of waves and their crashing on the rocks, the sigh and gurgling of the waters rushing through the wave channels, the hiss of the little wavelets dropping into the tide pools, were all about David in his babyhood, and it was inevitable that at last one of these waves would sweep him into a tide pool. Down he went under the water, sputtering and scrambling to get his nose out. But in his efforts, he found himself suddenly swimming in the two-foot-deep pool. His mother, who had left him asleep on the beach under the protective shelter of a rock while she went out to sea to feed and was gone longer than she figured, came back just in time to watch him proudly as he swam across the pool and then pulled himself up on the other side. She nosed him higher on the rock shelf, anxious that he not get too far into the waves, as she knew the danger for little seals.

But once David had passed the first few weeks of just sleeping and feeding and growing that prepare a pup for a more active life, he became adventurous. Soon he was finding his way higher onto the beach, where other pups had gathered to play and sleep above the harem areas and away from the danger of rampaging bulls. Here his mother could leave him alone more safely for a while, as she fished in the rolling swells of the blue green sea. And here he learned to push and struggle and bite, in play of course, with other pups, but sometimes roughly enough to cause yelps of pain. All this was good muscle-growing exercise for young sea lions who must soon face the dangers of the sea.

When older pups began to move down to the tide pools on the edge of the sea to practice swimming, David followed, and it was sheer luck that his mother caught him in time, confused and alone, swimming out into the foam-flecked waters, resolutely rising on the crests of the waves and sliding down into their hollows. She led him quickly back to the safety of the shore, but even as she did so, three other lost pups in the sea bleated in terror. Sharp rows of teeth in open white maws rose beside them and snapped at their bodies. Without their mothers they had no defense against sharks, the tigers of the sea.

Yet it was through swimming that David had to learn the art of escaping these and other enemies, as well as catching his food, in the years to come. So she made him understand, by swimming with him and herding him back before he reached deep water, that at present only the shallow areas were safe enough for him to try his flippers. It was the great front flippers, each almost a third of his total length, that he used at first in swimming, sculling through the water like an oarsman in a race, learning to drive suddenly forward with increased speed when a comber curled over him, so as to miss the churning tearing action of the falling wave. But as he practiced navigating the surge channels between the rocks, and got some bangs on his sides when he misjudged the waveswept currents, he began also to catch and feel the wonderful power and control that came when body and mind and flippers worked together in unison as he twisted and thrashed through the waters. Gradually he learned that at greater speeds he could use his hindflippers more effectively than his front, when they worked together with his sinuously twisting body. At such times the foreflippers acted more like steering oars.

By the time the great bulls had left Año Nuevo in August and only the cows and pups remained on the beaches, David was everywhere, swimming with the swiftest and strongest youngsters, playing tag and hide-and-go-seek and king-of-the-castle in the big tide pools and on the fringes of the sea where the waves screamed over the rocks. How brave then to dare the largest of all the young seals of that year to pull you off a steep rock surrounded by sea, and to push and shove and snarl and bark until, exhausted at last, you were pulled off your kingdom by three or four other young ones, and down with an inglorious splash into the sea!

Unlike the fur seals, whose mothers leave them to find their own way lonely and frightened into the far seas at about four months, David's mother stayed with him, watching, teaching, leading, until at last, confident of his powers to face the open ocean and its dangers under her guidance, she took him away from the island in late October. Most of the time they followed the kelp beds up the coast, moving from bed to bed of these large floating islands of giant seaweeds, where the mother felt safer with her young. Her senses were alert for sharks or killer whales, so that she could get into the deep shelter of the beds where her youngster could swim through the mazes of the stems where no large animal could go.

After many trials and disappointments, for learning to hunt is not easy for any young animal, David finally caught his first fish. The satisfactory snap of his jaw closing on halibut or rockfish gave him a new joy. But for this he had to learn how to turn and twist as swiftly as his prey moved through the waters, follow it down and down sometimes into the depths, like a wolf on the trail of a wounded deer. In such chases he learned how to hold his breath properly and to use all the oxygen stored in his body in muscle, vein, artery, blubber and lungs, while depressing the beat of his heart to allow as little of the precious gas as possible to be lost. Some of this came to him by instinct, but some had to be learned.

He could stay below for only two minutes at first, then lengthened the time day by day and dive by dive until he could do half of his mother's long twenty minutes without breathing below the sea's surface.

Now they were north of San Francisco Bay in colder waters off the Sonoma coast, and the fear of the big sharks was less, but of killer whales more. Deep in the waters one day, when they were fishing together, mother and son were making the sharp clicking sounds that some scientists believe are a somewhat cruder form of sonar than that used by porpoises for echo-locating prey, enemies or obstructions. Perhaps they were really clicking to locate fish or squid below them in the darker layers of the sea, but suddenly from above they heard a deeper clicking sound and an occasional trill or whistle. David sensed his mother's instant alarm and turned to follow her as she headed for the surface, slanting away from the direction of the sounds.

When they surfaced, they took deep breaths and began to swim at full speed for the nearest shore. They could hear the thunder of the surf ahead, but behind them they could hear that ominous clicking coming closer and closer. At full speed they swam, about fifteen miles an hour, every nerve and muscle concentrated on swift movement, torpedoing through the waves rather than over them, catching their breaths in every hollow between two waves.

If David had looked back, he would have seen a great black fin cresting the swell behind him, a fin that is the warning sign of the coming of death from the sea, the flag of a three-ton, twenty-eight-foot killer whale male, intent on his fleeing prey and moving at better than thirty miles an hour. The mother seal, sensing a desperate maneuver was in order, suddenly changed direction, shooting off at nearly a right angle, with David making the turn perfectly beside her. It was an angle the speeding killer with his great bulk could not turn into, but the swishing reach of his flippers nearly touched them as he shot by and swirled to get at them again. But a half minute of time was gained by this change of direction, and the two sea lions dove between two large rocks near the shore, rocks the enraged killer missed from smashing into by an inch. Then mother and son flung themselves exhausted on the rocks of the shore and were safe!

There were some greater hunters than the killer whales, who stalked David late the following spring off the Washington coast. He was nearly a year old, still getting milk from his mother, still learning from her the tricks of the sea, when she suddenly left him for a deeper dive than he could manage, going far into the dark waters after squid. He waited for her near a coastal rock about a hundred yards from shore, where there was a narrow shelf on which he could climb if a shark or killer whale came too close.

Unfortunately for him, he had been observed through a telescope from out to sea, and a fishing launch headed his way at high speed, keeping the rock between it and the young sea lion. The droning sound of the motor, heard only vaguely around the rock, did not frighten him, and he did not know he was in danger until the launch, engines shut off, and moving only by its forward motion in the sea, circled around the rock and suddenly was upon him. As suddenly, a large net on a pole was shoved out and over him, dipping down and catching him as he tried to dive. He struggled valiantly to free himself, biting at the net, but his two hundred pounds of weight was handled easily by four strong men, until they could get him to the side of the boat and inject his thrashing body with a tranquilizer. When he woke, he was aboard the launch and lashed to the deck with cords he had no chance of breaking.

It was fortunate his captors were biologists, and his next home was a large cement tank surrounded with a walkway, a basking platform and high walls, kept fresh with seawater brought through pipes from the sea. It was not a sea aquarium where people could watch him, but a place prepared by scientists for observations of seals. In the area with him were two female California sea lion yearlings, a female and a male of his own species, both about a year older, and two harbor seal females.

The year he spent in this prison was not a happy one, after his glorious year as a free creature in the sea. He was taught to obey commands by being given rewards of frozen fish, or sometimes by punishment when he did not obey, and he was also subjected to many tests to try his abilities and observe his reactions when swimming, diving, catching food, and his possible use of sonar in fishing. The biologists meant to be kind and never deliberately harmed him, but it was uncomfortable at best to be tied tight in a sling, lowered into the water and held under to see how long he could hold his breath. Of course, when he began to struggle violently to get out, they quickly brought him up to air, marking down in their books the number of minutes and seconds he had stayed there. He did not appreciate their exclamations of joy the day they got him to stay under a full twenty minutes! Worst of all was being put into a dark tank of water and there subjected to pressures equal to that of the sea two or three hundred feet below the surface. During and after each such test all kinds of gadgets were attached to his body to register body temperature, blood pressure, oxygen depletion and anything else the scientists could think of!

In the swimming pool environment itself, David was subjected to something still more unpleasant, the domination and bullying of the other and older Northern sea lion male. This domineering character, whom we will call Goliath because he was unusually large for his age, considered the younger and smaller sea lion fair game for constant yelling at and attacking. Goliath was particularly pleased with himself because his bark was gradually acquiring the deep roar of the adult male, and he exercised his voice at every opportunity by shouting at David. If David did not get away quickly from any area where Goliath did not want him, particularly near the female seals, he would chase him and get a bite in if he could. David was quicker at travelling and at dodging, so he did not often get bitten, but he greatly resented getting practically no time in the pool but always being chased out of it. Even when he retreated to the high basking platform when no one else was using it, Goliath sometimes drove him from this place too just for the devil of it!

Of course this male sea lion dominance over a younger animal of the same species is natural in sea lion society, and was watched with great interest by the biologists, who noted and recorded every move and sound, but it was far worse for David in the confined space with no escape as he would have had in the wild. Some of the marks of Goliath's teeth he would bear for most of his life, and he remembered these encounters and his humiliation for a long time.

Freedom came most unexpectedly. Actually the scientists had been training all the seals in the enclosure eventually to be taken out by boat and observed swimming free in the sea. All had been trained to return immediately for reward to their trainer at the sound of a special whistle, but to make sure they would return when called, or at least could be captured again, each was fitted with a harness to which an empty rubber bag was connected. This bag would be inflated by a device set off by an automatic timer, and once inflated, would prevent the seal from diving and slow his speed through the sea, so he could easily be run down by boat and captured.

When the animals were let loose in the sea Goliath chased and attacked David. In attacking the surprised David, Goliath bit through part of the harness and cut it free, while David, fighting back, broke the timer on Goliath's harness, so that it failed to operate and blow up the bag. The result was that some chagrined scientists got only four animals safely back to their boat, and two escaped!

David, once free, made every effort to leave his enemy as far behind as possible, and Goliath, content with having got in the last bite, soon went off in another direction, both animals diving and disappearing when the men in the boat tried to chase them down. David celebrated his miraculous escape by swimming in close to a rocky shore with deep water nearby and diving for rock cod about two hundred feet down. The flash of silver fins, the wild flight in chase, the satisfying crunch of his jaws when he finally swerved and caught the dodging fish, was like sweet medicine to his long-wounded spirit.

That summer David returned to the rocky beaches of Año Nuevo Island as a bachelor seal, and, with other bachelor seals, found a part of the beach where there were no great bulls with their harems, and the bachelors were left alone to play and fight and sleep. Near their beach they occasionally saw a huge elephant seal haul out on the sand to rest and sleep awhile. Once David crept close enough to poke a somnolent giant with his nose, but the huge beast snoozed on in such complete oblivion that David even crawled over his body without being noticed! He would have been surprised to come to the same island in winter and see the elephant seals at their breeding grounds, with hundreds of fighting, bellowing and loud-clapping bulls and thousands of roaring cows. The two species, elephant seal and sea lion, have entirely different breeding and nursing periods, perhaps nature's way of preventing friction between two such different animals.

Goliath also was among the bachelors that summer, as a three-year-old, but David carefully avoided him, keeping his eye on him and remembering vividly the times Goliath had treated him as an inferior. The other bachelors nearer his own age and David had many a mock battle, going through all the motions of the great bulls, lifting up their heads with the sign of pride and strength, trying to roar like them, and lunging and snapping at each other, but without the savage intent to do great harm.

So the years passed, until David was six years old and dove for the first time six hundred feet beneath the surface of the sea to drag a large octopus from its hole among the undersea cliffs. He tore and cut it with his long canine teeth, while the octopus struck back with its tentacles and razor-sharp beak, until the long sharp teeth of the sea lion reached the brain of the octopus and David's prey was dead. Then he dragged it to the surface, trailing blood, and was chomping down his meal when he found himself suddenly surrounded by three ten-foot sharks, lured by the smell of blood. But now David was ten feet long himself, 1,800 pounds in weight, and no longer a small pup lost in the waters, easy shark bait. He turned in three flashing movements and drove the sharks away, his roar of anger smashing through the waters.

These victories gave him such an elated feeling that that summer he headed south for Año Nuevo Island and the breeding beaches, knowing there would be other huge bulls there with their harems, whom he might now challenge. If you had watched him as he swam, you would have seen a bull sea lion more slender than normal, but also more swift and supple. He enjoyed cresting the uplifting seas to spy about or to porpoise out of the water in a great curve that caught the sunlight on his wet and golden back in shimmering beams. And when he dove for fish or squid his powerful body was the epitome of slick frictionless curves, every part synchronized to slide through the water with maximum precision, down to the dark depths where his large eyes could catch the faint luminescence that marked his prey. Then he would turn and twist with every zig or zag of a fleeing creature till his jaws snapped on yielding flesh. Of course there were many times he missed and a fish that could dodge fast enough flashed away to safety. As an old experienced hand at deep-sea fishing, he scornfully rejected the dark inky pseudomorphs, imitations of themselves, that the squids jetted back to distract his attack, but drove around the squidlike black clouds to seize the actual prey and kill with one powerful cruch of jaws and teeth.

When he returned to Año Nuevo, he heard again the long-remembered roaring of the great bulls, echoing the roar of the waves of the sea. He came in to shore through a rocky channel where the green seas surged his body high, so he could look over the rookeries and see who was present. There, on the most favored site on the beach, he recognized Goliath as the greatest of the bulls, a full 2,250 pounds in weight and over thirteen feet long, a monster swaying high and roaring above his harem, to whom none from sea or land gave answer. David's heart rose as he thought that the only bull he cared to fight was Goliath; but the monster's huge size caused that heart also to fail him, and he turned away to the bachelor beaches to train himself for another year.

So, by his own feelings, he was kept from the females for another season, though in time all among the bachelors bowed before his strength and quickness. He and the other outcasts from the harems tried out their battle tricks, not striking with full force but with enough to make muscles ache and heads spin. He learned a good trick one time when a larger bull challenged him. He turned and galloped off suddenly as if in fright, the other hot on his trail; then suddenly stopped and swung in a tight circle to catch him with a well-timed blow from the side that sent the other reeling, committed as he was to going straight forward.

David swam north that July to the cold northern seas as far as the Aleutian Islands, and then north again to the Pribilofs where the fur seals gather in the hundreds of thousands. He even swam on to icy St. Lawrence Island, where the last of the walruses were following the retreating ice packs into the Arctic seas. There he was nearly pierced by the tusks of a great bull walrus, short-tempered because David had swum too close to him out of curiosity. The two-ton monster turned suddenly on him with lightning speed, lifting those three-foot tusks and bringing them down like white streaks, and David rolled sideways just in time, but still felt one of them knife down along his skin leaving a streak of blood!

This long journey and its many adventures, out to the end of the sea lions' world, strengthened his muscles and gave him sea wisdom beyond his age, for death was always waiting for those not quick and strong. That winter and following spring he fished among the Shumagin Islands with his distant kin, the fur-seal bulls. He rode the wild storms of the North Pacific down to the Inside Passage of the continent's northwest wall, learning to let the great seas of storm throw his body far up toward the dark clouds, but riding down the waves and into their hollows with all the aplomb and skill of a champion surfer. And when the warm days came again, his blood, hot with male desire, called him to Año Nuevo and the ancient sea lion rookeries, where fate waited in the form of the long dangerous teeth and mighty body of Goliath, greatest bull of the beaches.

David came up cautiously through the rocks and the long tidal channel between them that led to the harem beaches. Riding the top of a long green swell of the sea, his heart beating more rapidly than usual, his eyes alert, he suddenly saw Goliath reigning over his harem of more than twenty cow sea lions. He saw his massive form, his long white teeth, his proud and imposing head; but this time his heart did not fail him. Close to a ton in weight himself, the hot blood flowing through his arteries and veins and the muscles gathering power, he moved up onto the great rock shelf and roared his challenge.

When Goliath saw him coming, the monster reared up two-thirds of his more than thirteen feet in the air and roared with a voice that seemed to shake the earth, and all about fell silent, watching. As David reached and reared up beside him, the two pointed their heads to the sky and roared in unison, but the challenger was a full foot and a half shorter than the master of the beaches. Who could doubt who would be winner?

Then Goliath struck, and all his mighty muscles and weight went into the blow, his great teeth like curving scimitars, aiming for the neck and perhaps the jugular vein. But David, the supple one, had moved too rapidly sideways for the blow to touch him, and as he dodged, he struck a glancing blow that brought blood flowing from the mighty neck. In mad rage, Goliath fell upon him, rending, tearing and shaking, so that David would have been killed in a few more seconds had he not used every steel muscle to tear himself free from that grip, leaving bloody flesh behind, and then turned to run. Only his speed saved him, for Goliath rushed savagely after him. Two pups were crushed like gull eggs, and a 600-pound cow was flung ten feet through the air as she tried to save her baby. The bulls paid no more attention to her than if she had been a rock crab.

Just when Goliath thought he could close with and tear to pieces his rash opponent, David swerved his supple body swiftly in a tight circle, and came from the side as the great body flashed by. He smashed with his ton's weight of bone and muscle at the root of Goliath's fore-flipper, his memories of shame and pain going into the furious blow, and ripped and tore with his teeth as his power knocked the giant sideways. There was nothing Goliath could do. His momentum in one direction left him no force to take this sudden blow from the side. As he fell and felt the flipper ripped down the middle, he knew he was now outclassed in combat, and limped with numbing pain. With a last bellow of frustrated rage, he threw himself down the sloping rock beach and dove into the surge channel, where the outgoing tide would carry him to sea. His action in fleeing proclaimed to all who saw him that there was a new conqueror on the beaches!

California Sea Lion

[Zalophus californianus]

Description The male is up to eight feet in length, weighing as much as 1,000 pounds; the female up to six feet, weighing up to 600 pounds. The male is black to dark brown, the female usually light brown or tannish, shining when wet. There is no underfur. The adult male has a distinctive bulge or crest on the forehead, unlike the Northern sea lion, which has a sloping head. Ears are quite tiny, but visible. The flippers and their use are described in the last paragraph of this section.

Range, habitat and some habits Ranges from about the 49th parallel off the British Columbia coast south to the central Mexican coast, but breeds in summertime from the Santa Barbara Channel Islands south, usually on rocky islets or isolated beaches. The rest of the year it is found on many rocky beaches and sea rocks up and down the coast, but most numerously from San Francisco south. It likes open rocks, unlike the Southern fur seal, which prefers caves and crevices. During the breeding season the dominant bulls form harems of several cows in a regular territory on the rocks, and here the pups from the previous yearly mating are born. After breeding, the males move gradually

California Sea Lion [*Zalophus californianus*]

north, often as far as central British Columbia, while most females stay with their pups near the breeding areas, where the water is warmer.

Locomotion and food   The California sea lion is often the performing seal at circuses and zoos or large sea aquariums, where its extraordinary dexterity and its ease in being trained makes it able to perform feats of leaping and diving, balancing balls, and so forth, to the delight of audiences. The long foreflippers are used for sculling the water in unison (sometimes described as flying underwater) at slow and medium speeds; at high speed the whole rear part of the body combines with the hindflippers to drive the sea lion at speeds up to twenty miles an hour, using a rapid twisting motion. On land the hindflippers are drawn or rotated forward, allowing the animal to climb rocks with some agility and even to gallop over flat sand nearly as fast as a running man, the forepart of the body rocking from side to side as it runs. This species eats mostly noncommercial fishes and squid, caught while driving forward at full speed underwater; also some molluscs and crustaceans. Its life is very similar to that of the Northern sea lion as described in the previous section.

FAMILY

ODOBENIDAE:

Walrus

Only one species, as described below.

Walrus

[Odobenida rosmarus]

Description Males are ten to twelve feet long, females seven to eight feet; the former weighing up to one and a half tons or more, the females up to one ton. Both have very thick bodies, with several inches of blubber protect-

Walrus [*Obobenida rosmarus*]

79

ing them against cold. Heads are rather small in proportion to bodies, with huge white ivory tusks protruding, especially large and straight in males, fourteen to forty inches long; curved and smaller in females. The grayish colored body is almost hairless, the skin quite wrinkled, the nose very blunt. The toes of the front flippers have hard nails, of use in digging out clams at the bottom of shallow seas, as are the tusks. Two flat and three pointed nails are found on the hindflippers, which can be bent forward for locomotion on land or rough ice.

Range, habitats and some habits   On the Pacific Coast walruses are found from the Pribilof Islands in Bering Sea, north to Bering Straits during most of the year, but many migrate in summer north through Bering Straits into the Chukchi Sea and Arctic Ocean. Because of their heavy weight they prefer thick ice as a habitat, with strong enough sea currents nearby to make open leads of black water, but they can live beneath the ice by digging holes up to the air with their tusks. They are preyed upon by polar bears and killer whales, as well as men, particularly Eskimos, who use all parts of their bodies, flesh for food, blubber to burn for light and heat, skin for boat coverings, intestines for window "glass," and so on. Walruses use their tusks with extraordinary dexterity, and their ability to cooperate to defend themselves as social units make them very dangerous to attack, even by the largest polar bears and killer whales. These prefer to attack the young, sick or wounded walruses, especially when separated from the herd, rather than the healthy adults. (For more details on this see the polar bear chapter.) They have been known to attack men in small boats, wrecking the boats and killing the crews.

Locomotion, young and feeding behavior   Walruses are comparatively slow swimmers, about four to five miles an hour, but faster than polar bears. They use their front flippers for sculling and their long hindflippers as propellers. Walrus calves are born on thick ice or ice floes from mid-April to mid-May, at which time they are about two feet long, growing to above five feet, the size of a harbor seal, by mid-August. While taken into the water at an early date by their mothers, they at first cannot swim, but cling with flippers tightly to their mother's neck, and are only gradually taught to swim. They are watched, fed and carefully guarded and protected by their mothers for about two years. Females do not breed again until at least one year after the birth of their calves. The mothers form nursery herds for mutual protection. Walruses feed mainly on molluscs, particularly clams, they find and grub out from the mud with their tusks. Some rogue males become hunters of smaller seals.

Harbor, or Spotted, Seal [*Phoca vitulina*]

PHOCIDAE:

Earless, or Hair, Seals

No outer ears can be seen on these seals, though they have middle and inner ears, and a tiny hole to the surface of the skin allows them to hear. The hindflippers are mainly for use in swimming and cannot be rotated up beside the body for fast land travel or climbing rocks as can those of the eared seals. Adult seals have a characteristic stiff fur with no underfur, but in some species the young are born with lanugo, a white woolly fur, for early protection against cold. The teeth show no difference between molars and premolars in the adults; the incisors are simple in shape with pointed crowns.

Harbor, or Spotted, Seal

[*Phoca vitulina*]

Description Up to six feet in length, weighing as adults from seventy to three hundred pounds; both sexes of about the same size. Usually grayish in color with white spots, or yellowish gray with black, brown or gray blotches; no external ears. It has no underfur, and hair is very coarse. Hindflippers stay stretched out behind body, which is better for swimming, but severely limits ability to climb rocks or travel fast on land.

Range and habitats   Found mainly in harbors and bays along the coast from northern Baja California to Alaska, but in Bering Sea is more an inhabitant of open sea; sometimes goes up rivers. Small herds are often visible near McNeil Island in Puget Sound, near Tacoma, Washington, and on a sandbar near the east end of the San Francisco Bay Bridge at low tide. In the Bering Sea in early spring harbor seals give birth to pups and raise them on the loose ice at the southern rim of the pack ice, but farther south pups are usually born and raised on beaches.

Locomotion and food   Best speed swimming about fifteen miles per hour. Like other earless seals, the harbor seal uses its permanently extended backward hindflippers to propel it through the water, also moving the body sinuously. Feeds mainly on fish, squid, crabs, shrimp, and other marine life.

Twin Life Stories

PRELUDE

Some scientists believe that the hair or earless seals of the family Phocidae, to which the harbor seal belongs, were originally almost entirely ice dwellers, raising their pups in an environment of ice and snow. The most evident proof of this is that the fetus of all these seals is usually covered just before birth with a prenatal white fur, which remains among those seals that continue to rear their young on the ice, but disappears before birth in those pups that are born in warmer areas. One theory is that this white fur is not only to protect the newborn pups from the cold in icy areas, but also to camouflage them by making them appear like part of the snow.

It is highly interesting that the harbor seal shows both phases of this evolution in its wide range on our coast. In the north, in Bering Sea, the females seek out in early spring the southern fringes of the pack ice, where it is most broken up, and give birth to and rear their pups on these loose pieces of ice. The pups are born with the white woolly fur which those born further south lack. Being born on loose ice is a fairly good protection against polar bears, who fear to swim out so far to sea, but it is little protection against men, who come in high-powered motorboats with rifles. However, neither the flesh nor skins of harbor seals is very valuable; besides, they have quick reflexes and can get out of sight when men come.

Where harbor seals are found close to civilization they are subject to many diseases and also to poisoning from the pollution of harbor and bay waters, so that often these usually beautiful animals have scabby skins or blinded eyes. Fortunately even the blind are still able to fish by their use of echo-location in the murky waters.

Selective habitats of ice-loving seals   To understand the specialized habitats and some of the associated habits of all the ice-loving seals, it is best to compare them in this special section. It is very interesting that each kind of hair or earless seal, and also their relative the walrus, in the Bering and Chukchi Seas, finds an area during the raising of its young to which it is especially adapted. This special adaptation eliminates much competition from similar kinds of seals. As noted above, the harbor seals rear their pups in the most unstable ice of all, that nearest the southern fringe of the spring ice pack. The ribbon seal also likes unstable ice for its pups, but in a somewhat different form, and is generally found farther north in the Bering Sea on denser and thicker ice floes. Bearded seals, on the other hand, are much larger and the thicker blubber under their skins gives them greater resistance to cold. They like to breed in the thicker ice still farther north, even as far as the Chukchi Sea. They need, however, strong sea currents to produce open leads of water between the larger ice floes. Their pups plunge into this open water four or five days after birth, protected by enough blubber by then to stand the cold water, and so avoid those polar bears and men who are able to walk out onto this thicker ice. The walrus also likes very thick ice, but its powerful and dangerous tusks, and its social instinct of moving in herds, make it unnecessary to get its young into the water immediately. In fact young walruses do not know how to swim at first and may be carried in the flippers of their mothers, or cling by their own flippers to their mothers' necks.

The ringed seals breed nearest to the land masses and on the most stable ice of all—that which is anchored to the land. They like this kind of ice because around the pressure ridges, and near the anchored icebergs, there are thick snowdrifts under which they can dig and build snug little dens, completely hidden from polar bears and other predators, though not always hidden from the predators' keen sense of smell. The best of these dens are made in deep crevices in the ice, where a very narrow entrance may keep a polar bear out. These dens during the cold of early spring protect the cubs not only from predators but also from the icy Arctic winds, and for a far longer period than any other seal pups enjoy, often five to seven weeks. The ringed seals are the smallest of all the seals in the Far North, and a rule of the Arctic is that the bigger you are the better protected against cold you are, so the small ringed seal pups need the extra protection of family closeness. Polar bears hunt out the dens of such pups by travelling over the ice and snow to where they believe the dens are, carefully swinging their noses back and forth as they go over the surface of the snow, until they scent a den below, then digging down to it!

A HARBOR SEAL IN THE NORTH

One pup was born in early April on a slab of ice about twenty-five by sixteen feet in size and about ten inches thick, which floated at the southern edge of the great Arctic ice pack in the middle of the Bering Sea. This ice slab, like many others of its kind in the same area, was the kind of unstable temporary base for raising a seal pup that the mother harbor seal wanted, for she was choosing an area where polar bears rarely dared to venture. But she also wanted her pup to be out of the water for the first three to four weeks of its life, so it would not be attacked by killer whales or other large denizens of the sea.

The pup gave a shrill cry as soon as it was born, and the mother moved silently to lick and groom it, then drag it along a way to break the umbilical cord. It was born with the white lanugo fur that protected its small body from the icy cold but would not stand much wetting, so the mother kept it in the middle of the small ice floe where it would remain dry even under the weak spring sunlight, which came from a sun just a few degrees above the horizon. Itti-vik we will call him, an Eskimo name meaning something like "small sea spirit." Soon she was feeding Itti-vik milk to fill him out with blubber until he no longer needed the white fur, which would be replaced by more water-adaptive hair. When he was full and happy, Itti-vik fell asleep on the ice and his mother left him to hunt for fish in the sea, slipping under water silently and gliding like a pale ghost through the icy green waters where small and medium-sized fish darted before her. The pup dozed for a few hours and then woke to give his shrill cry of hunger, a cry that might have brought a predator. But even the killer whales did not like this area of broken cutting chunks of ice, too dangerous to their enormous dorsal fins.

"Maa-aw!" Itti-vik cried shrilly again and again, and his mother came up from the sea in a graceful leap, sliding her body over the ice toward him. Again and as always she was completely silent, for after being weaned from milk, all spotted or harbor seals lose their voices. She gave him her milk and lay beside him to sleep, her light-colored yellowish and spotted fur merging with the yellowish white ice. He also slept deeply, his first few weeks of life being made up mainly of eating and sleeping.

Here and there on similar chunks of ice other mother harbor seals were also suckling their white-furred youngsters or sleeping beside them or diving into the water to hunt for food. Nowhere at this time was there any sign of males, for these happy-go-lucky creatures were off in some other areas, fishing or playing, completely free of any family responsibility.

When Itti-vik was a little over four weeks old, he had grown enough blubber under his skin to protect him from the cold Arctic winds and the icy cold sea, and began to lose his white fur. Now his mother led him into the water where he began to swim with as much ease as a newborn human baby begins to breathe. He also began to play in the water with other seal pups and even make futile attempts to catch fish, while his mother brought him chewed-up pieces of fish from the depths, weaning him gradually from his earlier milk diet. She herself was soon fertilized by a wandering male, who came casually out of the blue green sea to mate with her in the water and then as casually vanished into the depths. In her few days of being in heat several like him came to mate with her promiscuously, as harbor seals have no harems.

In a few more weeks Itti-vik was weaned and swimming in the sea rapidly enough to begin to catch his own fish. In the time-honored excitement of the chase, he followed his mother in the dive and the search through green-glinting waters, with the clicking sound of his new and inexperienced sonar trying to locate the movements of bodies before him. Shoals of small fish fled like streaks of light ahead of him, and he wildly drove after them, his flippers stroking the water behind him like powerful sculling oars. Finally came the all-out rush of power to close in on a fleeing prey, and the snap of his jaws. Sometimes they closed on nothing, for the fish could dodge and he was yet young and inexperienced; sometimes they connected satisfactorily with flesh. The fish knew their enemy; they darted and wove through the water in scintillating patterns, dodged often, and some always escaped, those fastest and most agile, or wisest in the ways of flight.

The ice was breaking up now, retreating northward with the advancing heat of the May and June suns. The mothers and their pups swam north with the retreating ice, for the ice floes still furnished protection from sharks or killer whales from the sea or an occasional rare but deadly bull walrus who had turned outlaw and seal killer. With them traveled a small group of the rarer ribbon seals and their pups, whose vivid yellow rings around gray bodies shone in the sunlight.

And now one day there came a strange sound out of the west, merely strange to Itti-vik, who leaped out of the water as high as he could to see where it came from. But the sound filled his mother with fear, for she knew it meant danger, the coming of man.

A seal hunter's launch came creaming over the waves at a speed far faster than seals could swim, the three men in its bow armed with high-powered rifles. The near-by ringed seals leaped in haste onto a large ice floe and gathered at its middle. Resting there they watched in stupefaction the launch approach, believing the strange object could not get up on the ice, and that they were safe. At the sudden explosions of the rifles, three ringed seals fell beneath the impact of the bullets, their pups crying piteously. But the harbor seals, far more wary of men, had already dived under the largest ice floes they could find and were swimming away in all directions as if killer whales were after them! Itti-vik and his mother swam rapidly underwater, coming up on the farther side of an ice floe to catch some deep breaths, then diving again and heading for another floe far away from the men and their loud explosions.

As the months passed Itti-vik grew toward adulthood and began the long yearly rhythm of the life of the northern ice-loving harbor or spotted seal. As a yearling in the following fall he traveled south with some of his kind, exploring the southern limits of the Bering Sea, and met other harbor seals along the Alaska peninsula who were afraid of the ice and sought warm beaches in summer to raise their young. These seals made no sounds either, but talked with a wave of the flipper, a toss of the head, a slap upon the water, the touching of noses. Otherwise their habits were different from his and their actions strange to him and he did not stay with them long.

Still later, as a three-year-old, he travelled in summer with some old males, far north beyond the Bering Strait and into the Chukchi Sea, a branch of the Arctic Ocean, and there he saw herds of walruses thick on the deep ice islands and feared their bellowing and their flashing tusks. His fear vanished when he found he could swim faster than any walrus, and in late fall he even flowed south with them in their great migration to the less chilly waters of Bering Sea where they had their wintering grounds. In his fourth year he mated with several female seals soon after their pups were born on the spring ice at the southern fringe of the great northern ice pack in Bering Sea. And so his bloodline was passed on to future generations, the line of a seal people long acquainted with and very wary of man, and so perhaps likely to survive through many generations to come.

PUGET SOUND PUP

The other pup was a female, born far away from the cold ice of the north, on a warm summer beach at the very tip of Puget Sound on Gertrude Island, Pierce County. It is one of the safest places in the world for seals, because it is under the protection of the great gray walls of McNeil Island Federal Penitentiary, near Tacoma, a prison run by the United States Department of Justice, near which no boats are allowed.

The tremendous variability and adaptability of harbor seals is shown by the fact that this birth took place in the third week of September, four and a half months later than that of the pup born on the ice of the Bering Sea. This wide difference in time is partly explained by the good fishing in southern Puget Sound in early fall, which gives a new pup plenty of nourishment, while in the Bering Sea it is May before the ice and sunlight are just right for mother seals and their new pups. During the middle and late part of May pups are born on the ocean shores and nearby bays of the Pacific coast of Washington, while in the northern Puget Sound region new harbor seal pups appear from the end of July through August. In California pupping time is from March to April, much the same as the early time in Bering Sea.

In spite of the protection of the federal penitentiary to the herd of harbor seals on nearby Gertrude Island, all was not well for them there. The newborn female pup's mother was blind, one of four blind mothers on the island, and this, along with skin diseases and other infirmities common among the Gertrude Island harbor seals, was caused by the impurity of the water in Puget Sound from city sewages, paper-pulp mills and other sources of pollution. It was not the clean and beautiful green blue waters of the central Bering Sea, but a murky liquid in which seals could only see a few yards at best. Even with good eyes the harbor seals would have had difficulty hunting for the fish, squid, molluscs, crustaceans and other sea life on which they feed, but fortunately they have another aid to hunting, without which the blind mother would long ago have starved to death. We do not yet know the details, but it may be the same clicking of the teeth used by some sea lions, by which the harbor seals send out series of sharp sounds that act like sonar, echo-locating fish and other prey in the waters so that the seal can zero in and seize them without the use of eyes.

The newborn harbor seal pup, whom we shall call Lina, from the species' scientific name *Phoca vitulina*, was born without the white lanugo fur that had covered the northern pup at birth. In the warm Puget Sound region it was not needed. But what was needed was the affection and closeness of Lina's mother, and when her mother was with her, Lina was content to feed deeply on the warm milk and then sleep on a grassy area above the beach. This area was separate from the main herd and the mother seals had decided it was safe for their pups. When the pup awoke, however, and found her mother gone, she would lift her muzzle to the sky and utter a sheeplike "Ma-a-a," or if that did not bring fast results, a more urgent and plaintive "Kru-oo-ruh!" The mother always approached in that strange silent way of harbor seals, uttering no sound herself, but coming close and nuzzling Lina with her nose, then scratching and caressing her all over while the baby fed.

Very soon Lina was led down to the water, where the gentle waves of the Sound were safe for small seals, and where she took to the water as naturally as she had to the milk, swimming beside her mother, playing with her and even riding on her back. On the shore the mother usually tried to stay above Lina, especially when a boat came near the beach and she sensed possible danger. This enabled her, when necessary, to drive her pup into the water, where she knew the two of them could dive quickly and be safer from man's attack. However, if the mother heard the distant sounds of motorboats, the shouts of young people, and the sharp cracking of twenty-two-caliber rifles, she dove deeply and led her baby to a place immediately under the walls of the great prison, where she had learned from past experience the young hunters were not allowed to go. Perhaps she would have been pleased if she had heard and understood the deeper sound of a police motorboat followed by the arrest of those indiscriminately shooting seals. Many a seal, wounded by those bullets, has fled to a hiding place among the rocks, there to die after hours and perhaps days of suffering from the thoughtless acts of such people. They enjoy shooting at wild creatures, but have no concept of their feelings and their pain, probably quite equal to those of humans!

When the nursing period was over and Lina was weaned, about five and a half weeks after her birth, she began to leave her mother to play with other pups on the edge of the herd. She also began seriously to seek for her own fish and molluscs and crab and shrimp food in the cloudy waters and among the tide pools of the shore. Though clumsy and often failing in her efforts at first, it was astonishing how quickly she picked up the hunting practices of the older seals, especially her mother's, learning by watching and listening to improve her ability to hunt and catch.

Perhaps her greatest lesson came when the great black dorsal fins of a pack of killer whales clove the waters of Puget Sound near the island. The old seals instantly slapped the water loudly with their front flippers, and every seal within hearing headed for the land. Lina was swimming two hundred yards from shore when the flipper-slapping was started. She grew confused because her mother usually slapped water as a warning to get off land and into water. This slapping was done in such a way that she sensed it meant something different. She started into deeper water, then, as she came up to breathe, she saw the foam trails of many seals heading for land. Too late she turned to swim after them. There was a vast swirling motion near her and a mouth as big as an oven opened to show two rows of fierce white teeth. With ruthless efficiency it started to close on her, when an ultimate reflex, imbedded in her from generations who had been hunted by these same giants, made her twist her body sharply at right angles to the snapping jaws. Down and away she went, hearing the sharp clicking noise of the killer whale's sonar trying to echo-locate her and the other fleeing seals in the murky waters. She swam through a foaming mass of blood where two other seals had met their death only seconds before, and made desperately for shore. Again she sensed a great swirling body coming up from behind her, and again she jackknifed superbly in such a sharp turn that the larger creature could not follow her, and she missed the shearing teeth again by inches. In a few seconds she reached rocks near the shore and dove between them, through a gap too narrow for a killer whale to follow.

Out of such experiences, including the sharp sting one day of a twenty-two-caliber bullet cutting through the fleshy part of her neck, Lina built her life-style avoiding dangers. One subtle danger she learned to avoid through her sense of smell, more highly developed than her mother's, for she shunned all smells in the waters of the Sound that bore the hint of pollution. She thus missed the dubious distinction of being blinded by pollution as had her mother. After she had mated and borne her own baby seal, she tried to teach him this wisdom too. Perhaps she and other seals like her will live one day to see and smell cleaner waters in northwestern America!

Ribbon Seal

[Phoca fasciata]

Description   Both male and female are about the same size, around five to six feet long; weight up to 200 pounds. This is a small brown to grayish seal marked with distinctive yellow bands around hips, front flippers and neck, but the females are much less brightly colored than the males. Fur stiff in adults and without underfur, but young have white woolly underfur for a time after birth.

Range, habitats and some habits   Found mainly from the Aleutian Islands north in Bering Sea, a few passing in summer northward into the Chukchi Sea by way of Bering Strait; a few others moving south in winter along the coast as far as central California. Usually pelagic, staying away from land; in the summer and fall being most common in the Bering Sea, but rare elsewhere. In winter ribbon seals stay near the southern edge of the pack ice in Bering Sea, and in spring females rear their pups in the broken ice of the southern fringe of the ice pack, but slightly farther north than the harbor seals who like the extreme southern edge where there is more open water. The pups are born on the ice, usually in early April, and are covered with white lanugo or natal fur, which protects them from cold before they have grown sufficient blubber under their skin, and may also camouflage them from seal-hunting polar bears. The pups do not enter the water to swim until such fur is shed, about three to four weeks after birth when enough blubber has grown under their skins to let them stand the icy waters. Pups are weaned in about a month,

Ribbon Seal [*Phoca fasciata*]

and learn first to eat fish caught by their mothers and then to catch their own.

Locomotion and food These small seals are capable of swimming up to ten miles an hour or more, using front flippers for steering and hindflippers and whole body in sinuous motion for propulsion. They feed on fish, molluscs and crustaceans, the latter two food items being used mainly by the young.

Ringed Seal

[Phoca hispida]

Description The smallest of our hair seals, being usually a little less than five feet long, and both sexes weighing up to 150 pounds. Color is dull brownish to yellowish with dark streaks and spots usually in continuous lines along the back. Pale buff rings appear along the sides, but these are not very conspicuous.

Range, habitats and habits Found throughout much of the Arctic regions; in our area mainly concentrated in Bering

Sea from which it migrates north in summer to the Chukchi Sea and the Arctic Ocean. This seal is the chief food of the polar bear during the winter and spring, but, because of its habit of living around land-fast ice, it is also preyed upon extensively by foxes, wolverines, wolves, dogs and man. It is little bothered by killer whales because of its close association with thick ice. During winter it lives mainly under the ice, gnawing out blowholes where it comes up for air (see story of polar bear). In spring females give birth to pups in special dens they have hollowed out under deep-packed snow on the land ice, from which a tunnel usually leads through into the sea for escape and food gathering. Dens are also sometimes made in crevices in broken rough shore ice, which are probably safer from polar bears than those under packed snow, which a polar bear can dig through when he smells a seal pup. Young and inexperienced mothers may hide their pups in dens in the more open pack ice away from land, where they are more easily found by the bears. The pups are born covered with white lanugo fur, which has no camouflage value since they are always born and kept hidden for several weeks deep beneath the snow. The mother may leave her pup alone in the

Ringed Seal [*Phoca hispida*]

den for some time while she goes fishing. The dens, of course, are much warmer for baby seals than open pack ice, and the small size of ringed seals makes it necessary to keep them there. The pups are weaned after five to seven weeks, usually in May, and then taken into the sea wearing the new swimming hair that has replaced the lanugo. These seals love to sun themselves on the ice floes during the summer, taking short catnaps, but ever watchful for polar bears or other enemies.

Locomotion and food   Swimming is done mainly by twisting the whole body sinuously, the hindflippers always stretched out behind and operating as propellers; speed about ten miles an hour, or more when frightened. Feeds mainly on small Arctic fish, but also molluscs and crustaceans.

Bearded Seal

[Erignathus barbatus]

Description   Length of adults eight to eleven feet, weight up to 800 pounds or more, sometimes even half a ton, the largest of all the earless seals on our coast; uniformly yellowish or gray in color, but unique in having very prominent tufts of long flattened whiskers or bristles on each side of the muzzle.

Range, habitats and habits   Usually these huge seals travel alone. Occasionally they gather in numbers in particularly favorable food areas, but never form distinct herds. They range from the Bering Sea northward into the Arctic Ocean. Pups are born in the spring on fairly thick ice, north of the broken ice of the southern edge of the pack ice, but away from land-anchored ice favored by the ringed seals, usually between April 5 and May 3. The pups are born with thick soft gray (or sometimes silvery bluish to brown) hair, but the white lanugo fur found in other ice-loving earless seals is shed before birth. The large size of the pups at birth, and the quick growth of blubber from the rich milk, as well as their fur, allow the pups to enter water within three to four days after birth, where they are safer from polar bears. For this reason the mothers select places to give birth that are near open leads of black water caused by strong sea currents. During their first few weeks in the water the mothers watch over the pups carefully and guide them. These seals are too large and powerful to be bothered by foxes or even wolves on the ice. The pups weigh about seventy-five to eighty pounds at birth, but are nearly four feet long by the time of weaning, about fifteen to eighteen days after birth—an astonishingly short nursing period for such a large animal. At the time of weaning a pup may weigh as much as 190 pounds, much of it made up of blubber, which forms a fine insulation against cold.

Bearded Seal [*Erignathus barbatus*]

This is extraordinarily different from walrus calves, which weigh about as much at birth but are nursed and cared for by the mothers for from eighteen to twenty-four months.

Locomotion and food   The bearded seal is a swift and powerful swimmer, capable of speeds up to fifteen miles an hour or more. As with other earless seals, the hindflippers are permanently stretched behind the body, which twists sinuously in harmony with the hindflippers to create rapid propulsion. The food is mainly fish, but with some eating of crustaceans and molluscs. Large quantities of fish must be eaten to maintain body health. (For more about bearded seals see also the chapter about the polar bear.)

ORDER

CETACEA

(Whales, Dolphins, and Porpoises)

The mammals of this order include some of the most spectacularly large of any creatures in the history of the world. No dinosaur, large though some of them were (up to thirty to thirty-five tons), came anywhere near the blue whale, who is sometimes as much as 105 feet long and may weigh as much as 150 tons! The only creature alive today near in size to whales is the harmless plankton-eating whale shark (Rhincodon typhus), which may be forty to sixty feet in length, and weighs from fifteen to twenty-five or more tons.

Almost all Cetacea are highly adapted to living in the sea, but a few porpoises are found in rivers such as the Ganges and the Amazon, the Ganges porpoise being so specially adapted to the silt-clogged murky waters that it is born blind! Though having a fishlike body, the whale shows its difference by possessing a tail with horizontal instead of vertical flukes; and strange to say this "tail" is not derived from the original boned tails of land mammals, which the whale ancestors were, but is entirely made of flesh, skin and muscle!

The proof that whales and their relatives were once land mammals is that the bones of hind legs are found hidden in the bodies of most whales, at least in their fetuses, but they no longer have any function. Another strange fact about whales is that they have become so adapted to sea life, with their need to move through water with minimum friction, that they have lost all their hair, their external ears, and any sign of external genitalia. The blubber under their skin, sometimes up to fifteen inches or more thick, protects them from cold water, and their size also helps them maintain a high body temperature in the ice-filled waters of the Arctic and Antarctic.

The two main suborders of the Cetacea are the toothed whales of the suborder Odontoceti (including the immense sperm whales and the considerably smaller dolphins and porpoises and their relatives), and the baleen or whalebone whales of the Mysticeti (including the largest of all whales, the blue whale), which have no teeth. The former prey on squid, fishes and, sometimes, other whales, while the whalebone whales sift small marine creatures from the sea through specially designed sieves made of whalebone.

Whales apparently changed from land mammals to sea mammals at least as far back as the Middle Eocene epoch (around forty-five million years ago) when we have fossil evidence of the first appearance of an animal something like the modern toothed whales. This is many million years before the first seallike animals left the land and took to the seas. Even in these early whalelike creatures the nostrils had changed from the front of the face, as in all modern land mammals, to the top of the head, which allows for easier breathing when swimming (something seals have never developed). How such a radical evolution from a land mammal to a sea mammal actually took place, we have only slight ideas. But we do know that some modern fish that live on the bottom of the sea by lying sidewise have one eye migrate during a lifetime from the side of the head to the top where it can see upward. Similarly the nostrils of the Cetacea probably migrated through myriads of centuries from the natural nose position of a land mammal to the top of the head. The Mysticeti or whalebone whales are a more specialized type and did not begin to appear in fossil form until the Miocene epoch (about twenty million years ago), possibly from a small-toothed ancestor of the Oligocene (30 million years ago).

SUBORDER

ODONTOCETI

(Toothed Whales)

These are the only whales with teeth, though some do not show their teeth until they are several years old; others may have the teeth buried in the gums.

ZIPHIDAE:

Beaked Whales

These whales are called beaked because of the long snouts that look like birds' beaks. They also are characterized by a rather small, sometimes sickle-shaped dorsal fin, almost two-thirds of the way back from the nose end. They usually have one large tooth (occasionally two) on each side of the lower jaw, always present in males, but sometimes hidden in the gums in females. These teeth are probably useful in catching food, but the scars on the bodies of the males also show they are used in fighting between rivals.

These are the most mysterious of all whales, and very little is known about them. They appear to be most closely related to the sperm whales and have a similar method of gathering food, diving deeply after cuttlefish, squid and octopus. Some scientists believe that some of the large Ziphids, such as Baird's bottle-nosed whale, may dive even more deeply than the sperm whale, as much as 3,500 feet under the sea! But there is no sure evidence of this that I have heard of, and they do their diving without the aid of the giant sperm-oil-filled head of the sperm whale, which is believed to assist it in its *deep dives. Also, unlike the sperm whale, the female beaked whale is generally a bit larger than the male. Some beaked whales are reported to have harem-type schools, each dominated by a harem bull. In this they again resemble the sperm whale bull, although the latter is much greater in size than the female of the species.*

Baird's Bottle-nosed, or Beaked, Whale

[Berardius bairdii]

Description   Length of both sexes up to forty feet or more. The black back and sides often appear to be covered by a number of white scratches on adults, probably caused by fighting during mating, or even from fights with large squids, which these whales capture from deep under the sea. The undersides are dark grayish to whitish. The dorsal fin, unlike that of the other beaked whales in our area, is not concave on its back side. The forehead rises almost straight up from the back of the beak and is quite well defined as compared to other beaked whales. The side fins are not pointed but blunt.

Range and habitat   The range of this rare and little known whale probably covers most of the north Pacific Ocean and the Bering Sea.

Locomotion and feeding   A fairly swift swimmer, but I can find no records of speed. It feeds mainly on squid up to a fairly large size, and there is evidence they dive very

Baird's Bottle-nosed, or Beaked, Whale [*Berardius bairdii*]

deeply for them, even as deeply as the sperm whale, or down to 3,500 feet below sea level. Fierce fights with the well-armed large squid probably occur at these levels.

One thing that has prevented study of these whales has been the small interest whalers take in catching them because they lack much whale oil. Thus scientists on whaling ships have had little chance to observe them, and unfortunately scientific expeditions at sea have also neglected studying them, probably because of this same lack of economic importance. Yet it is highly possible that they are among the most interesting animals of the ocean.

North Pacific Beaked Whale

[Mesoplodon stejnegeri]

Description   Up to about eighteen feet long, rarely to twenty. Black all over except for white beak and part of head, but body often marked by white scars, as on other beaked whales (see previous entry). Two five- to eight-inch-high teeth with prominent tips are found in lower jaw of the male. The small dorsal fin has a slight concave surface on rear side.

Range and habitat   Apparently a very rare denizen of the north Pacific Ocean in the open sea.

Habits   Feeds on squid, as the hard beaks of squid are found in the stomachs of stranded specimens, but probably also feeds on octopus and cuttlefish.

TOP: Goose-beaked Whale [*Ziphius cavirostris*]; BOTTOM: North Pacific Beaked Whale [*Mesoplodon stejnegeri*]

Japanese Beaked Whale

[Mesoplodon ginkodens]

Description   Similar to the above, but dark gray above and light gray or whitish below.

Range and habitat   Open sea from Japan to Bering Sea.

Habits   Probably feeds on squid, cuttlefish and octopus.

Goose-beaked Whale

[Ziphius cavirostris]

Description   Length up to twenty-eight feet, but average for male about twenty-two feet, for female twenty-three feet. Variable color but usually some mixture of blackish, purplish or lead gray with white. Head and upper back usually white; a peculiar whitish ridge extends down back from the dorsal fin, which is unusually large and narrow for a beaked whale. Throat grooves make large V under chin. Male has two round teeth, about two inches long, projecting from front of lower jaw. Numerous white scars tell of conflicts with squid or with other whales of same kind.

Range and habitat   Travels in occasional schools of up to about forty or more off Pacific coast from Baja California to Bering Sea.

Habits   Probably feeds on squids, octopus and cuttlefish. Can dive under the sea for thirty minutes or more.

Sperm Whale, or Cachelot [*Physeter catodon*]

PHYSETERIDAE:

Sperm and
Pygmy Sperm Whales

The skulls of these whales are shaped asymmetrically, or out of proportion on one side, with the blowhole single and on one side. Inside the skull is a large organ or space for carrying sperm oil, much larger in proportion to size in the sperm whale than in the pygmy sperm whale. The lower jaw is usually underhung.

Sperm Whale, or Cachelot

[Physeter catodon]

Description   Adult males reach lengths of fifty to sixty-five feet, females twenty-five to thirty-five feet. The remarkable difference in size between sexes is often a sign among mammals of a harem-type family life. Both sexes of adults are dark bluish gray in color, but some may be whitish on the belly and lower jaw. The lower jaw has twenty to twenty-five large teeth, each up to eight inches long in adults, but sometimes worn down to almost nothing in age. These teeth fit into sockets in the upper jaw (which has no teeth), and are used mainly for catching prey or cutting large prey to pieces, but not for chewing it, as most food is swallowed whole. In bulls the teeth are largely used for fighting other bulls. The huge and remarkable head, which contains a reservoir of oil, probably of use in diving, is one-third of the whole length of the whale, and has only one blowhole, which appears on the left side, and which spouts at about forty-five degrees in a forward direction. There are humps on the rear back, but no fins. Side fins are comparatively small for a whale.

Range and habitat   Found in oceans all over the world, but the females with young rarely go north or south of 45-degrees latitude, while males range into Arctic and Antarctic waters in summer. Females give birth to their calves in the tropics, as warmth is necessary for the thin-skinned babies. On our Pacific Coast these whales rarely come within sight of land except where the water is quite deep.

Food and locomotion   Food is mostly squids, with the octopus perhaps second, the cuttlefish third, and various kinds of ocean fish fourth. The females and young feed mainly on middle-sized creatures and down to depths of perhaps a thousand feet, but the large males make herculean dives lasting even an hour or more, and down to depths as much as 3,500 feet or more. There they tackle the largest squid and octopi, some of the giant squids being sixty feet or more in length and weighing several tons. The beaks of squids decompose in a peculiar way in the rectums of sperm whales and provide a remarkable and extremely valuable substance called ambergris, used to make expensive perfumes. It has been suggested by one scientist that the smell of ambergris may aid the sperm whale in attracting squid.

Sperm whales generally travel at rates from three to five miles an hour, blowing at intervals ranging from every thirty or forty seconds to as long apart as five minutes. When frightened or enraged speeds of ten to fifteen miles an hour or even more have been observed.

A Life Story

PRELUDE

Sperm whales have been reported harpooned by Phoenician seamen in the eastern Mediterranean as long ago as a thousand years before the time of Christ. If so, it must have been a heroic feat for men in a small ship of those times, and an event to be long told about in song and story. Down through the centuries sperm whale hunting, especially of the bulls, has probably been mankind's most exciting and dangerous chase of a wild animal. This is because the sperm whale is the largest of all the toothed whales and because it is capable of using its huge head more effectively as a battering ram than any other whale, sometimes actually shattering small whaling ships, such as in the famous sinking of the *Essex* off the coast of Peru in 1820. As recently as 1964 a small but ocean-going yacht was attacked, apparently without provocation, in the South Pacific near the Marquesas Islands, and nearly sank.

Although a sperm whale was a formidable antagonist to whaling boats of the old days, many of which were sunk by these whales, modern oceangoing ships are, for the most part, too big and too strong to be in any danger. Primitive peoples, such as the Makah Indians of northwest Washington, the Eskimos, and the early Basques of northern Spain were experts at attacking and killing smaller and more inoffensive whales, but apparently they left the far more dangerous sperm whales completely alone.

The slaughter of whales by our modern whaling seagoing factories and their catcher boats has been responsible for the deaths of tens of thousands of sperm whales. Sperm whales have had one advantage over the huge baleen (or whalebone) whales, such as the blue whale and the humpbacked whale, in that, because they are toothed whales, they can find the food they need in most parts of the oceans. The baleen whales, especially during the summer, find their plankton food or krill in certain concentrated areas of the northernmost and southernmost parts of the oceans where whaling ships can easily find them. The sperm whales, spreading out more widely in the vastness of all the oceans, have a better chance of avoiding their major enemy, man, and so have not been so nearly decimated as the baleen whales.

Despite their fearsome record, it is amazing how inoffensive sperm whales can be toward man if they are left alone. They even seem to sense when men mean no harm to them. *Calypso,* the famed oceanic exploration ship of Jacques Cousteau, and its accompanying smaller boats often passed through herds of sperm whales in perfect peace, with the divers able to go down to observe them.

Sperm whales seem to show some of the same foibles as human beings, our strengths and our weaknesses. Sometimes they can be very brave and ferocious against an enemy, even a ship larger than themselves, sometimes very noble in rushing to the aid of a wounded or threatened comrade. Other times, even the largest of them can tremble in fear at the threat of something unknown, rush away in wild terror, or even became so paralyzed with dread that they lie still and let themselves be killed. They also like many other kinds of whales have a strange obsession at times to throw themselves up on a beach to die. Sometimes some disease they have has destroyed their ability to detect danger, but at other times they appear in the best of health but seem to have given up all hope and merely want to end their lives. Could they have committed suicide because of a lost love or the death of a beloved one? We do know that whales are very sensitive creatures, far more affected by each other and by moods than are most other mammals.

One of the great mysteries of the sperm whale is its large reservoir of spermaceti or sperm oil, which accounts for much of the enormous size of its head. Some scientists have surmised that this foamlike oil is of vital aid to the whale in its deep-sea dives, when it goes more deeply down into the ocean than possibly any other mammal, where the water pressure is almost incredible. Other matters that are still mysteries revolve about its social life, its means of communication, and its ways of getting food.

CACH AND THE HAREM

The female sperm whale had been growing a fetus within her belly for fourteen and a half months, one of the longest gestation periods known to any form of life. The gestation periods of whales are geared closely to a year in length to fit their migrations between the feeding grounds of summer and the calving times of early winter. But the sperm whale follows no set migration pattern, finding its home almost anywhere in the oceans and as likely to be in the tropics as in the northern or southern seas during the summer. The cow had come with the harem herd to which she belonged in late August to the comparatively warm waters three hundred miles off the coast of the southern tip of Baja California, and here in a calm sea on the last day of the month she gave birth to her baby. He was quite a baby—he weighed more than a ton and was about fifteen feet in length!

Unlike human babies he already knew instinctively how to swim, but his mother did have to nudge him to the surface of the water to make sure he could get his first breath, and teach him to come there whenever he needed a new one. He enjoyed a distinct advantage over most mammals in having a comparatively large brain, well convoluted like a man's and capable of learning much as he grew older.

A large shark that had got too curious and excited by the blood in the water from the umbilical cord had to be struck at and driven away by the mother's tail flukes, and then chased still farther away by a nurse whale, a female without a child of her own, who had come up to help the mother. Only then did the mother relax and turn on her side to present her nipples to the newborn. At first he was clumsy and several hours were spent in teaching him to turn sideways too and put his mouth near one of the nipples which protruded from one of two slits on the whale's belly far behind the navel. He did not suck as the calf of a land cow does, as his undershot jaw was not adapted to it, but when his mouth touched the nipple, his mother expertly shot a stream of milk into his mouth in a creamy jet. One-third of it was composed of fat, as compared to one-twentieth of a land cow's milk. This was designed to add rapidly to his bulk, especially to the layer of blubber under his skin to protect him from cold weather and from the attacks of sharks. The warm milk spurting down his gullet made him shiver with delight, and he flexed his body to seek more of it.

All around the baby was a harem herd of about fifty whales, most of them females with young, but with one large old male, sixty feet in length, and two smaller males of forty-five- and fifty-foot size. It was not like a harem of fur seals, where one male jealously dominates his group of females, allowing no other bulls to enter his territory. The giant old male was the main harem bull, but he allowed his younger and smaller companion males to be part of the herd, as long as they did not directly challenge his authority or try to take first choice among the cows. It was somewhat like the social system of a wolf pack in the Canadian wild, but probably without quite as sharp lines of dominance and control.

On that day and in that part of the ocean, almost as far as the eye could see, whales spouting or showing their great dark backs could be seen, with the longest and highest spout about a half-mile off from the main part of the herd, and each spout going up at the 45-degree angle characteristic of the sperm whales. As the long black bodies reared and lowered themselves in the waves the bulls could be distinguished by the several small humps on their hindbacks, while the cows had generally only one. Below the water surface ultrasound equipment could detect sounds much more important to sperm whales than the evidence of their eyes. High sharp clacking noises are the sonar sounds whales send out to make their echo-locations. Deeper grunting sounds are those the whales make to communicate with each other.

The calf, whom we shall call Cach (short for Cachelot) knew nothing about these sounds at birth, but he soon began to hear them about him in the clear green waters of the deep sea, and sensed they meant something of importance to his mother. As the days passed he began to imitate them awkwardly, just as he was clumsily imitating his mother's movements in swimming.

He had just begun to swim like a real sperm whale, moving his tail flukes in the ancient rhythmic way of his kind, when suddenly the whole herd of cachelots began to swim with faster than regular speed toward the north. From far away the great bull leader had picked up the distant sound of a large pair of propellers driving a ship through the Pacific. Unpleasant experiences a few years back, of which he was reminded by the occasional pain of a deep old harpoon wound in his back, had taught him this sound meant danger.

The faster than usual movement would have been too much for Cach if his mother had not dropped below him and swum in such a way that the currents created by her great bulk carried him along with her, the two swimming as one! A few miles at this higher speed of about eight to ten miles an hour got the herd away from the sound of the propellers, but brought them near another ominous sound from ahead, the sharp clicking echo-locating sounds of a pack of killer whales.

Unlike some other whales of our Pacific Coast, such as the gray whale or little piked whale, who would possibly have fled from this sound in fright, or frozen into fearful silence, the huge bull sperm whale directed his herd on, right toward the killer whale pack. The cows in the herd, however, protective of their calves, moved closer together, cows without calves also moving close to the mothers to give the calves double protection.

The killer whales, soon aware of the presence of the three great bull sperm whales, and so many others, and knowing the power of a whole herd, moved aside respectfully and passed on with only a few slight feints at a calf or two at the back of the herd that they hoped to separate from the mothers. If a calf had been severely injured and there was no hope for its life the lead bull would have ordered its abandonment for the sake of the whole herd.

In a few weeks Cach was playing with other calves, especially when his mother sounded, which means throwing her flukes up straight to the sky and diving deeply into the ocean to search for squid and other food. During such times he and the other young ones were watched and guarded by a nurse cow, who would circle anxiously around the play area, much as a den mother hovers over an active den of cub scouts. The games of chase-and-be-chased were probably as old as the mammal kingdom, but often the young and exuberant sperm whales made up new games such as throwing a small drifting log out of the water with a sharp toss of the head to see how far it would go, or sneaking up on a seabird resting or sleeping on the waves and giving its legs a nudge with the nose. It was very satisfying to hear it squawk protestingly as it rose from the surface.

Sometimes there would come a command from the

great herd bull, and the other whales would listen intently, wondering what he had heard, and a faint but deep grunting noise might come back, brought as all deep enough noises are perhaps dozens of miles across the sea from distant sperm whales sending out queries through the green waters. The herd bull would reply with a very low note like a bass drum beating and this would go speeding away under the waves in answer. In a few hours they might hear the beating of giant flukes, as another herd of sperm whales loomed up through the green sea shimmer to join them for a while and exchange information about the far seas or the terrible doings of men. Such double herds did not stay together long, however, for the law of the whale folk is not to attract crowds unless food in those parts of the sea is very plentiful indeed.

In December the herd had followed the warm currents southward, away from the advancing cold fingers of the north, for it was too early yet for the youngest whales of the group to meet the rigors of the northern seas. Off the Gulf of Tehauntepec they met a tropical storm. First came long lines of wavy clouds, then the sky darkened and the sunset turned a deep savage red. The sea tingled and all the sea creatures knew a big storm was coming. The small and the weak sought shelter, but the great and the strong were not disturbed. To Cach, however, as the storm grew stronger, the mounting waves and the shrill crying of the wind brought terror. Soon the waves were splashing into his blowhole and the wind blew needles of flying spume in his eyes when he surfaced, until he learned when the proper moment came to breathe and when to keep his head down, and how to rest in the trough between the waves for a few precious seconds. He saw some of the great whales coast down the steep slopes of the vast and roaring seas like boys with toboggans, dive underwater and come up again to smash their tail flukes on the wave tops, while the upper half of their bodies was lifted to the dark and howling sky.

Suddenly it was a new game and he was trying, with the other youngsters, to imitate these joyous elders, though sometimes the littler ones were lifted and turned over and over by a massing and breaking sea that threw even their three tons and more of weight about like twigs on a torrent. In the din and confusion of such a great wave breaking over him, Cach again became confused and frightened, strangling from the water in his blowhole, and trying desperately to reach his mother. Then suddenly she was beside him, having sensed his terror, a massive wall of flesh that comfortingly held back and tamed the raging sea!

When the sea calmed three nights later, the storm now moving northward, the calming water became filled with flashing lights, for a great school of albacore, thirty or forty thousand of the large silvery fish, were swirling through the dark but luminescent waters, stirring up the phosphorescent plankton as they chased smaller fish before them. Cach, startled and alarmed, rushed to his mother, but she paid no attention to him, for she was too busy catching fish. The albacore swam faster than she could, but she moved with them, dipping her head up and down and from side to side, even as she swam furiously, scooping one or more up now and then with the sweep of her lower jaw, even taking into her mouth and swallowing an eight-foot-long blue shark that got in her way in his own eagerness to catch albacore! Most of the fish escaped her by dodging, but in two hours she was full and content to loll near the surface, gradually drifting to sleep. Cach was beside her, ever so glad after such a frantic chase that now she had become still.

In the warm waters of the Pacific, just north of the Tropic of Cancer, night merged into day and day into night for Cach, in a life that brought something new every hour, mostly new sounds, all of intense interest to whales, who live more by sound than any other sense. Pistol shrimps cracked their pincers, fish of many kinds grunted, puffed or boomed from above and below, while seabirds could be heard calling shrilly when Cach lifted his head above the waves. Dolphins clicked and whistled and whinnied weirdly as they went racing by. His own kind used their clacking sonar and deep grunts to judge their environment and signal to each other. And all about him were the whispering, the sometimes crashing and roaring, the murmuring and singing of the mighty sea, never quiet, yet in its more peaceful times strangely soothing. Once he heard the deep humming, moaning and gurgling notes of humpback whales, the deepest singers of the seas, but as foreign and strange a language to a sperm whale, as the clicking, clacking talk of Kalahari Bushmen is to us.

What wisdom, untold and unknown to us, he must have learned during his first two years, which ended on the day he was weaned—a September day when his mother turned her back on him when he sought his familiar source of food, her milk. He was huge then by our standards, five tons or more in weight, but only a very tiny sperm whale among the immense bulks of his kind. Yet he felt as lost as all babies do when their milk supply is gone.

It was perhaps fortunate that five-hundred miles west of San Francisco, in the evening of that same day, with the western edge of the sky all suffused with dying pink, the clacking sonar of the sperm whales brought back a joyous message from below. Deep in the dark depths a vast cloud of squid was rising swiftly toward the surface, drawn by the coming of darkness and the shining first quarter of the moon. Now the echo-locating system of the herd moved into high gear, the sounds bouncing back from the deeps to tell them of approaching food, and they flipped their vast flukes skyward and dove almost as one.

Cach had gradually become used to diving, following his mother and other whales, each time a little deeper as he found he could hold the air, playing games with other whale youngsters to see who could dive deepest. So down he went now, after a series of breaths, down and down until he could see below him the luminescent white cloud that surrounded the rising squids. He liked the taste of these soft creatures whose tentacles writhed out at him and who tried to shoot away from his open mouth and teeth like fire-sprayed arrows going backward, or who made black clouds of ink to hide in, but could not escape him because he could locate them exactly with his clacking sonar. He opened his mouth again and again to swallow them singly and in clumps, as he shot gloriously here and there through their shining swarms, eating them as we might swallow spoonfuls of a delicious pudding. Of course, for every one he caught, a dozen or more squirted away from him to temporary safety.

There is a burden carried by a male born to the warrior breed of the sperm whales. It is a burden borne by the need for prestige, for fighting ability to win for oneself a harem of females. The largest and most mighty of living warriors, some as much as sixty tons in weight and over sixty feet long, each as large as twenty average-sized elephants, have such colossal destructive power that when two of them come together in mortal battle severe injury or death must follow to one or both. Since there is a general law of nature that it is bad for the species for the males to be too self-destructive, how can the sperm whales obey this law?

As a young adolescent male between the ages of about eight and fifteen Cach traveled most of the time with schools of other young males, after an earlier period spent in schools of mixed juveniles. The young females never formed their own schools, because they soon gravitated into one of the wandering harem herds. In his adolescent years in the different schools he joined, Cach matched his strength with other young bulls in what were largely mock battles and rarely gave great injury. They followed the ancient ceremonial rules of battle that sperm whales have followed for ages. Selecting a bull of about his own size or slightly smaller, he would send out his challenge by clashing his teeth together, the teeth of the lower jaw smashing up and into their receiving sockets in the upper jaw with a clashing sound heard for about half a mile or more at the age of eight, and around two miles when full-grown. These teeth make their appearance in the sperm whale's mouth only after about seven years, but by then they are ready for service!

The other young bull would rush to the surface, throw a third of his body out of water and swirl about in a circle, his dark eyes flashing anger until he saw his challenger cresting the waves. Then he would come down into the sea again, clacking his own teeth in answering challenge, and charge his challenger. Cach would also rush toward his enemy, the two approaching each other like large battering rams. But, like young bulls everywhere, they did not fully mean business. They slowed in the last seconds, collided with a jolt that sent echoes through the sea for many yards, and began battering and pushing each other with their heads, even occasionally locking jaws to pull and tug, but ready to stop when it hurt too much. The signal would then be given by one bull backing away, showing all the signs of wanting to break off the struggle. When these signals are properly given, the victor does not press his attack.

Once Cach saw a great herd bull, all sixty tons of him, rake a fifty-ton challenger all along one side with his teeth, as clean and sharp as a saber cut, the teeth ripping open the blubber six to eight inches deep and causing the blood to spurt. The power and expertness of the cut was so flawless and swift that the challenger fled without even the clashing of teeth, thoroughly realizing his inadequacy against such a veteran. Lives are saved in this way, and the greatest of the bulls reign for years with only a rare challenger, few daring to tackle them except the young and foolish.

It was in battles with giant octopi, cuttlefish, and squid deep under the sea that Cach built his own muscles, his breathing and his power, unknowingly training himself for a future challenge from a harem bull. His huge sperm-oil-filled head, his literal miles of veins, arteries, intestines and lung branchlets, his mighty yards of muscles and vast thicknesses of blubber, eighteen inches thick at the shoulder when he was sixteen years old, gave him the areas for oxygen storage and protection that made him able to dive sixty, seventy and even eighty minutes under the sea. Flipping his tail flukes twenty feet or more above the waves, when his clacking sonar sounds warned him of the presence of a gigantic living thing far below, he would point his body vertically downward after a series of deep breaths, and head for the dark depths of the sea.

Down he went through the first lighted areas of water, then into dimness and finally complete blackness, except for the scattered glittering lights of some of the strange living creatures of the deep sea. Now his sonar felt ahead of him as he dove deeper and deeper, the sounds echoing back to let him zero in on his prey. At 2,600 feet he shot suddenly ahead with an additional burst of speed, for he could see outlined below in faint luminescence a monster squid in whose translucent body the prey just eaten still glowed and gave away his position. The giant creature was jetting himself backward with powerful squirts of water toward a crevice in an underwater cliff as Cach closed with him swiftly. Sixty feet long was the monster, from arrow-shaped tail end to the tip of his foot-thick longest tentacles, and weighed several tons. As Cach rushed downward, the squid shot out a mass of inky black fluid that cut out all sight of him, luminescent contents and all, but Cach did not need eyes to know where his prey was. His sonar clacks were telling him the size of the giant and where he was.

Through the black inky fluid the great whale flashed, his lower jaw dropping and then smashing up toward the sockets of the upper jaw, as his mouth closed over the squid. The monster cephalopod wrapped his immense forty-foot tentacles about the whale's head, each three-inch-wide suction disk armed with sharp teeth that bit through skin and blubber. But most dangerous of all was the foot-long beak, sharp as a razor, that gouged into the eighteen-inch-thick blubber of Cach's shoulder. Cach felt no pain, only the lust of battle, his teeth shearing off tentacles as a bandsaw might rip through a log, and finally crushing into the squid's six-foot-long head and demolishing that largest and most intelligent brain of all the invertebrates, and ending its last bit of fight.

Cach cut his former opponent into convenient lengths and gulped some of them down, with a peculiar wriggling motion of his body that caused the folds of his throat to expand so he could swallow the enormous bites. Part of his meal eaten, and more than half of his diving time over, he seized the remains of the squid in his mouth, and headed back for the surface 2,700 feet above. Unlike a human diver, he was not bothered by the "bends," the decompression sickness of nitrogen in the blood that maims and sometimes kills humans, for his whole immense structure, including the massive oil-filled head, guarded him against such danger.

Though he came up swiftly, it was a long journey and he broke the surface in a great surge that threw him nearly completely out of the water. In the same instant he blew from the blowhole on top of his head a huge gust of fetid air, filled with drops of oil that made it appear as a cone-shaped geyser directed forward at his characteristic forty-five-degree angle. The whistling noise of the blow or spout echoed across the sea for half a mile or more. Sixty-five times he spouted, to let out the bad air and draw in the good, about one spout for every minute he had spent beneath the surface of the sea.

As Cach traveled about in the Pacific he reached most parts of its northern half, his farthest north being the Bering Straits between Alaska and Siberia. This is the northernmost reach of the ocean traveled by sperm whales, the males alone going so far, and then only when the warmth of the summer sun has opened up the ice packs enough to allow them breathing surfaces. Probably they do not go beyond the Bering Straits because the land masses there are too close together to let such huge creatures feel comfortable.

Sometimes Cach traveled with a harem herd, carefully avoiding the harem bull and of course not challenging him. He recognized his superior strength but felt him out through his sonar, and sometimes watched him as he repelled another bull. Once he saw a great harem bull tear off the whole lower jaw of a chellenger in one quick smashing attack. Cach was not ready yet to risk that. What taught him caution we cannot know, for many a young bull sperm whale, impelled by the blind ferocity of his maleness, attacks without thinking a veteran who can easily kill him in combat. Such bulls do not live to pass on their genes to future generations.

Other times Cach might travel for a few months with one of the great lone bulls of the north, bulls who had once been harem bulls but been defeated and often bore terrible scars till their dying day. Some of these have become the famous sperm whale bulls of history, such as New Zealand Tom, Timor Jack, Pata Tom, Mocha (Moby) Dick and others who sank whale boats and even ships. Possibly they had become so embittered by their enforced bachelorhood that they became like the mad bull elephants of Africa and India, who also were driven away from their herds. But these bulls seemed to appreciate Cach's companionship and he may have learned from watching and listening to them some of the ancient fighting wisdom of the oldest and greatest warriors of the sea.

So he came in his eighteenth year down out of the north, fifty-eight feet in length, and unscarred by battle, except for the marks on his head and shoulders made by giant squids with their cutting beaks, the long scars showing white against his dark hide. He came with the clacking sound of his sonar and his deep grunting call that could be heard for dozens of miles along the undersea valleys. His nearly sixty tons moved with power through the water, the vast flukes of his tail leaving a thick trail of churned white foam behind him, his colossal head bowing up at intervals of every three to five minutes to blow spume and suck in fresh air. In him the male urge and fury was building, yet somehow controlled by his cunning brain, so that when in February he located a harem herd of about forty whales in the far south tropic seas near the coast of Baja California, he made no challenge to the harem bull, no snapping of his teeth into the sockets of his upper jaw in the ancient sperm whale challenge. Instead he watched and listened for several weeks and felt and waited. The great herd bull, scarred victor of many battles, sounded a warning to him now and then, a short smashing sound of his powerful teeth, for he was worried. Cach paid no attention. He kept circling the herd, sometimes a mile away, sometimes closer, sometimes cutting through the middle of them, but avoided getting too close to the harem master.

The herd bull became increasingly nervous. He did not dare to dive too long after his usual food of giant squid, and grew hungry and irritable. Nor was he getting enough sleep. Perhaps Cach knew the right time to challenge would come, perhaps it was just accidental. The herd bull had finally taken a deep dive, driven by hunger, and come back to the surface exhausted from a battle in the depths with a monster squid. It took him seventy blows to get back his strength from his seventy-minute dive. Cach challenged him the minute the seventy were over and the huge bull lay resting in the waves. Ponderously toward him came Cach, clashing his teeth into their sockets, and the sound echoed through the water for two miles. At the deep, intimidating noise, the youngsters of the herd rushed for their mothers, while the adolescent bulls and females poised trembling in the sea, torn between fleeing and watching.

The harem master sounded. He lifted up his immense tail flukes and dove straight down into the sea, but only to about three hundred feet. As he surfaced he threw the whole front half of his body out of the water and turned about with the peculiar rotating motion of his flukes to look in all directions. On the surface of the water he could see the humped back of Cach moving toward him. As he came down he clashed his teeth into their sockets with a sound that startled the seabirds far around. One savage grunt, and he charged his challenger.

The two vast bodies careened toward each other at about twenty miles an hour, each just below the surface. At any instant the huge heads might smash into each other when an astonishing thing happened. Cach rolled sideways and away, so the other's head and body missed him. He rolled back again and took a nip that tore a chunk of flesh a yard wide from one fluke of the herd bull's tail. The water foamed red.

The herd bull then reversed the direction of his body with extraordinary agility, fired by incredible rage, and this time the two did crash together, Cach being unprepared for such a swift turn. He would have been swept under, and his side torn open, had he not himself, by an incredible effort of strength, pushed up far enough to come head to head with the older bull. The water clouded around them with blood and foam, teeth flashed and clamped, and they locked in that last terrible grip of combat, mouth across mouth, each struggling to tear loose or break the other's jaw. The immense bodies churned in the water, the flukes lashing till something had to give. Unlike the other bull's, Cach's tail was unharmed, and it turned the balance. Its leverage let him throw all his mighty muscles into one surge of power, to twist suddenly and break the herd bull's jaw. It was not a mortal wound, and the harem master tore himself loose with a mighty heave and, trailing blood, swam dazedly away. The jaw would heal, but it would take a long time, and the bull would never be the same again. No harem for him any more, but a lonely vigil in the northern seas, for a new king reigned!

Pygmy Sperm Whale

[Kogia breviceps]

Description   Length eight to thirteen feet; blackish above and grayish on sides and bottom; head is blunt and conical, almost bulbous in shape, but without the massive appearance of the sperm whale's head; blowhole a little bit to the left on head but farther back than in sperm whale; dorsal fin recurved like scimitar; lower jaw narrow and triangular in shape and partly recessed from snout. Flippers rather small; nine to fifteen very sharp slender teeth only in the lower jaw. Often covered with white scar marks on body, probably the result of fights with whales of same species.

Range and habitat   Rare in open sea from Washington southward to Baja California.

Habits   Most likely to be seen alone or in couples. Eats mainly small squid, but also has been found to eat crabs. Generally rather slow-moving.

TOP TO BOTTOM: White Whale, or Beluga [*Delphinapterus leucas*]; Narwhal [*Monodon monoceros*]; Pygmy Sperm Whale [*Kogia breviceps*]

Dwarf Sperm Whale

[Kogia simus]

Description   Similar to above, but smaller, six to eight feet in length.

Range and habitat   Very rare in temperate and tropical Pacific.

FAMILY

MONODONTIDAE:

White Whales and Narwhals

These whales are light in color and do not have a conspicuous back fin. The side fins are broad and paddle-shaped. Teeth are few or lacking.

White Whale, or Beluga

[Delphinapterus leucas]

Description   Length twelve to nineteen feet; adults are all white, but young animals are first dark gray, then become spotted with brown, gradually changing to yellowish before adult white. Snout very short with forehead rising steeply behind it. Neck visible.

Range and habitat   Found in Bering Sea and around the Arctic Ocean to northern Europe and Asia; likes rather shallow waters and leads between ice floes.

136

Habits　Rather slow swimmer, up to six miles an hour, living in small schools of five to ten. Sometimes swims up large rivers like the Yukon. Makes beautiful quavering and whistling sounds underwater. Probably feeds on slow-moving bottom fish, crabs, etc. Has proved tractable in captivity (New York Aquarium). A main enemy of the Beluga is the polar bear, who kills them when they become trapped at isolated holes in the ice. Belugas probably avoid killer whales by staying in ice pack areas the killers do not like because of their large dorsal fins.

Narwhal

[Monodon monoceros]

Description　Length (minus tusk) twelve to nineteen feet; single tusk of male six to nine feet long. General color white, usually with many small black dots or blotches, mainly on the upper parts. Young animals are bluish gray. Both sexes have two teeth in their upper jaws, though hidden by the gums in the female. In the male the left-sided tooth grows into a beautiful spiral tusk whose use is not quite clear, as it is somewhat delicate and can be easily broken and infected. It may be simply a male prestige item to impress the female.

Range and habitat　Likes shallow waters in Bering Sea and other cold waters of the polar region.

Habits　A rapid swimmer, in schools of five to twenty or sometimes many more; often only one sex in a school. May dive down to 1,300 feet for up to one-half hour. Whistles very shrilly when it surfaces. Female calls young with deep moaning. Feeds probably on small fish and crustaceans found at bottom of shallow seas.

FAMILY

DELPHINIDAE:

Small Whales (Dolphins, Porpoises) and Several Larger-toothed Whales

Most have conical teeth in both jaws; some, however, have teeth only in one. The two names, dolphin and porpoise, are rather confusing as often applied to the same species. Dolphin is the old Greek name and conveys a feeling of light and grace, whereas porpoise means "sea pig." By common usage most small whales of this family are called "dolphins" when they have beaks and "porpoises" when they have blunt heads, but even this distinction is not always made.

Common Dolphin

[Delphinus delphis]

Description   Up to eight feet long; very slender and graceful in appearance; black tinged with greenish on back, or
purplish brown; white or whitish below; golden stripes
along sides; black line extends from base of beak to fore-
flipper. Beak about five or six inches long; mouth with
forty-five to sixty-five pairs of small, sharply pointed,
conical teeth; a deep V-shaped groove marks head off from
beak.

Range and habitat   Abundant in the open ocean, but rare
near land, often seen in schools of hundreds or even thousands in most areas of the north Pacific except in colder
northern waters.

Habits   Schools often swim and leap together in perfect
harmony, almost flying out of the water in great flashing
curves, showing extreme joy in living. Speeds possibly up
to twenty-five miles an hour or more. Frequently ride in
the bow waves of ships, skillfully balancing on and using
the power of the wave with little personal effort. Many
yelping, squeaking and clicking noises are heard under
water. Common dolphins appear unusually sensitive and
are hard to tame or make happy in captivity, seem even
to die of broken hearts, perhaps longing too much for
their former freedom.

TOP TO BOTTOM: Common Dolphin [*Delphinus delphis*]; Pacific White-sided Dolphin [*Lagenorhynchus obliquidens*]; Striped Porpoise [*Stenella caeruleoalba*]; Northern Right-Whale Dolphin [*Lissodelphis borealis*]

Striped Porpoise

[Stenella caeruleoalba]

Once called the eastern Pacific subspecies of Stenella
styx, the gray porpoise, but here considered a separate
species. It is also closely related to Stenella longirostris,
the long-beaked dolphin or porpoise, which is more
evenly black above and white below, and which occurs
on the Mexican coast, but rarely in California waters.
A good classification of these species has not been made
yet and will take further investigation by specialists.

Description   Length up to ten feet; very beautiful in body
pattern, with back and head above black or blackish brown;
rest of body white except for two graceful and distinctive
black lines extending from the eye to the region of the vent,
and to the base of the flipper. The small conical teeth
number about forty-four to fifty in each jaw.

Range and habitat   Mainly found in good numbers from
Oregon north to British Columbia in open seas, but a
smaller number found south to the California coast and
north to Bering Sea.

Habits   Presumably they feed on fish and squid.

Eastern Pacific Spotted Dolphin

[Stenella graffmani]

Description   Up to eight feet long; has a very strange change in its color pattern during its life, starting with a dark gray saddle in the middle of the back, also extending to the head, while the underparts are light gray or whitish. This changes gradually with age, first with white spots appearing on the dark back while dark gray spots appear on the grayish white underparts. Then gradually the gray spots on the belly come together into solid gray with only a few white spots. Finally this dolphin becomes dark purplish gray or black above, with widespread white spots most numerous above the eye and on the sides halfway to the rear. The underparts become a fairly solid gray. Forty-three to forty-seven small conical teeth are in each jaw, roughly ridged and furrowed on top. Has a long and very prominent beak.

Range and habitat   In open sea from Mexico, north into southern California.

Habits   Travels in large schools, up to a thousand or more dolphins. Rapid swimmers, liking to ride the bow waves of ships. Feed mainly on fish and squid.

Pacific, or Gill's, Bottle-nosed Dolphin

[Tursiops gillii]

Sometimes called "cowfish"
or "common porpoise."

Description Up to twelve feet long, females a little smaller than males; weight up to 400 pounds or more; grayish or grayish purple on back, with blackish V-shaped mark on side of head; whitish below; flippers and flukes blackish; white splotch on upper side of nose. Eighteen to twenty-six pairs of sharp conical teeth in each jaw.

Range and habitats Found from central California (near San Francisco) south along coast, sometimes coming into bays or harbors and even rivers, but preferring more open sea.

Habits Like many members of the toothed whales, this dolphin has two layers of skin, one saturated with water and having special ridges that make the body so friction-less when passing through water that it can swim much faster than seems hydrodynamically possible. While this species is not the fastest of the delphinoids (dolphins, porpoises, etc.), it can do bursts of speed up to twenty-four miles an hour, and can maintain these high speeds with little effort by getting into balance when riding the bow waves of ships, so as to use their energy. These dolphins hunt in packs after fish, sometimes even surround the fish and drive them to each other; fish are also hunted at high speed by individual dolphins.

A Herd Life Story

PRELUDE

The Atlantic bottle-nosed dolphin (*Tursiops truncatus*) has been studied both in captivity and the wild, probably more extensively than any other of the whale order. The present species, though not as closely studied, is a very near relation to the Atlantic bottle-nose, and even may be merely a subspecies, so we can safely consider that its life activities are very close to those of its Atlantic relative.

The ancient Greeks and Romans, as well as the South Sea Islanders and other peoples, regarded the dolphin with both reverence and pleasure as being almost a brother to man. Many early stories tell of how they have carried people, especially children, on their backs, helped people in danger of drowning, and even assisted fishermen to catch their fish. In recent times many of these stories have been discounted as romantic fiction by scientists, until some actual examples of such events were proved beyond doubt in our own day. Thus, Opo, the famous dolphin of Oponi, New Zealand, played with children and even adults on the beaches there from 1955 to 1956, while Pelorus Jack, also of New Zealand, appeared to guide ships for many years (up to 1913) through the dangerous shoals between the North and South Islands. In our own United States and Canada, dolphins in captivity, and when taken after training into the open sea, have proven extraordinarily helpful and friendly toward man. Tuffy,

Pacific, or Gill's, Bottled-nosed Dolphin [*Tursiops gillii*]

the famed U. S. Navy dolphin of Point Magu, California, was even used to carry messages and lifesaving lines deep under the ocean to the aquanauts of Sea-Lab II.

Our recent interest in and experiments with dolphins so stimulated the imagination of many people that a few, such as Dr. John Lilly in his book *Man and Dolphin,* argued that the dolphin's large and well-convoluted brain, larger than that of man, made it almost certain that dolphins have a language and would eventually communicate with us. Further researches since this book have not yet proven Dr. Lilly right, and recently scientists have been very cautious about ascribing a true language to these animals, even though it is obvious that they have a high intelligence and a finer degree of sensitivity even than man's nearest animal relative, the chimpanzee. The difficulty with proving Dr. Lilly either right or wrong is that dolphins have few physical characteristics in common with us. They have neither hands nor a larynx capable of making varieties of sounds. This would make it very difficult to communicate with them directly, and isolating the sounds they do make and translating them into words seems next to impossible.

However, several promising discoveries make it at least probable that they have considerably more varied ways of communicating than any other animals. They do a lot of noise-making to communicate, as is clear to anybody who has listened to them through hydrophones at any of the large sea aquariums. Second, they have a high social organization and obviously work together cooperatively when fishing, repelling the attacks of sharks, or taking care of each other's youngsters. Third, they show very high intelligence in inventing games to play and in solving problems presented to them by humans, particularly in grasping a problem as a whole and solving it almost immediately, instead of making many false starts as most animals do. Fourth, it has been noted that some of them are of much higher intelligence than others of the same species, and some of these have actually looked and acted as if they were trying to communicate with us and then given up in disgust when we could not understand them! Fifth, they seem to sense in man a similar being, showing a kindness, gentleness and forbearance toward him that is often quite astonishing. For example, it is rare that a dolphin attacks a man even when treated badly by him or its young attacked. Sixth, there is growing evidence that the clicks, which scientists know are used for echolocation of food and enemies, are also quite possibly used in intercommunication by varying their pitch and the order, and the number and sequence in which clicks are given. If this is true, the variations could be quite enormous and would, when combined with other noises they make, such as whistles and squeaks, give the dolphins the capability of developing a language nearly like our own in complexity. Seventh, the killer whales, who are actually only specially large dolphins or porpoises, have only recently been studied in captivity, but are giving evidence of possessing intelligence superior to the smaller dolphins, an even higher form of social cooperation, and probably even more sophisticated use of sounds. Eighth, recent studies of dolphin communications show that dolphins are great imitators, sometimes even imitating the human voice; they may therefore be able to imitate the sounds their own sonar brings back to them from different objects. Their sonar not only senses size and shape, but also the texture and inner feel or quality of an object or creature. It follows that if they can repeat this exact formula to other dolphins, they are really using each such formula as an actual word for what they have sensed. Scientists point out, however, that using a word to name something, as a chimpanzee or a dolphin or even an intelligent dog sometimes may do, is still a long way from stringing words to create an idea. Most animals and birds can convey by a sound or even expressive movements the words "I am hungry," which is animal talk,

but none have yet been known to express even so simple
an idea as "I feel better when I sleep well," which is
using a language truly.

Maybe future research will demonstrate that dolphins
and killer whales can communicate such ideas, but we
still do not know. What we do know about dolphins as
living beings I have tried to tell in the following story of
the activities of a herd or pack of bottle-nosed dolphins
in the Pacific Ocean. Through watching the actions of different individuals through part of a year, we can get a
beginning picture of the complete dolphin life span.

DOLPHIN ADVENTURES

(*Note:* In the following story names for the different
dolphins are used as a matter of convenience, to sort out
some of the different characters. How dolphins may name
themselves we, of course, have no way of knowing. The
story of a herd, or pack, is given here because the dominant feature of the bottle-nosed dolphin is its social con-
sciousness and continual cooperation.)

In early July the bottle-nosed dolphin pack of about
fifteen individuals was cruising along the shore of California near Morro Bay, looking for schools of fish where
the sea was deep enough near the shore and where large
rocks rose from the waves. They were also expecting the
exciting and happy events of two new births. Tom, the
oldest, largest and wisest bull in the herd, the father and
grandfather of several of the others, was constantly on
guard against sharks. Even when the herd was chasing a
school of mackerel between some large rocks, he was
sending out the clicks of his sonar on a different and
lower frequency than the rest of the herd, who were
zeroing in on the fleeing fish. Tom's sonar was probing
for sharks and while he was doing this, he was also telling

his three oldest sons, Jack, Roger and Dark, to keep their eyes and sonar on guard too. He observed that Jack, the oldest, powerfully built and a good lieutenant, was beginning to send forth the same kinds of clicks and was swinging his head in different directions to shoot his cones of sonar through the water as a ship swings a searchlight on a dark night in a stormy sea. The two other sons were still intent on the fish, and Tom was less pleased with them.

Martha, one of Tom's three mates, a grandma of the herd and the most experienced of the females in directing a fish hunt, was serenely aware that her huge mate was staying aside from the chase on shark guard, and confident that he could handle any danger. At the same time, by her actions and the sound of her clicking sonar she showed the rest of the herd how best to corner and direct the fish for maximum catching, how to close in on the fleeing school of mackerel from all sides and drive them into a narrow place between the large rocks. And many fish were being gobbled up as the dolphins, by constant side-to-side motions, drove one fish into another dolphin's jaws. Martha meantime clicked impatiently at one of her daughters, Julie, who appeared lackadaisical about the chase. Then, by the action of Julie's body, she realized that her daughter was about to have her baby. With a series of sharp whistles, Martha called off the hunt and the herd milled about and let the fish go.

Tom had sensed that Julie was about to give birth earlier than most in the herd, and had been sending out low-frequency sonar signals to determine if any sharks were near. Now he whistled to his sons to form a barrier across one end of the channel between the large rocks, while Ralph and Tony, two still younger sons, equivalent to our teen-agers, were sent to guard the other end of the channel.

From the south end, where he now rested alertly in the gently rocking waves, Tom had noted earlier the presence of large sharks. He was afraid that when the baby was born, the sharks would detect with their extraordinarily keen sense of smell the odor of blood from the broken umbilical cord in the water. The four large bull dolphins faced south, watching and busily clicking away, while the females gathered about Julie with many clicks and squeaks and whistles. Her mother was specially close to her, ready to help with the birth.

Martha was happy to see the tail of the new baby begin to emerge as Julie heaved and struggled in labor, for if the head came first and the birth was even slightly delayed the baby might drown. All the whale folk females, unlike land mammals, are adapted to giving birth hind end first. It is the lighter end of the body in whales, as in land animals the forelegs and head are lighter. In both kinds of animals, as birth approaches, the heavy end of the fetus tends to sink lower into the belly, bringing the lighter end nearer to the opening through which the new baby will emerge.

As the baby dolphin slid slowly out, the entrance of the baby into its new world was aided by the softness and floppiness of both tail flukes and fins, which would become stiff and hard later, and the streamlined body slid out of the mother easily. Julie could not scream or groan like a human mother in the pain of labor, but all her attendants knew by the contraction of her muscles that she was enduring pain.

Almost the instant the baby was free of the mother, Martha began to push it toward the surface so it could get its first breath of air, and the somewhat dazed Julie also pushed the baby to the surface. The blowhole on top of the small head surfaced in the comparatively calm waters, and the little dolphin drew in its first breath. At the same instant the natal or umbilical cord from the baby's belly, through which it had drawn sustenance from the mother, broke in two at a weak spot, and spurted out a little cloud of blood into the water. Almost at the same moment, perhaps in sympathy, Susan, another young mother-to-be, started to give birth, and her mother, Cora, swam close to be ready to help.

The knowledge of the new birth and the blood in the waters was signalled to the four large bulls, and their sonar clicks increased in intensity as they swung their heads from side to side, spraying the sounds in different directions; for now the most critical time in the life of the pack was upon them. Fine and exact as the dolphins' sonar is in locating a fish or an enemy, finer still are the incredibly powerful smelling organs of sharks in locating the presence of blood. The scent of blood drove them to a frenzy of hunger and made them lose their usual respect for and fear of dolphins. So one by one the nearby sea sharks started moving toward the little bottle-nosed dolphin pack.

Tom's sonar was the first to detect the coming of a shark and he was relieved that it was a tiger shark and not one of the great white sharks that could, if large enough, almost decimate a group of porpoises or dolphins singlehanded. Against such a monster the only defense would be to find a narrow opening in the rocks and swim through it one by one. As the first hungry shark came swimming into sight, the dolphins' sonar was already picking up two or three more behind it.

Tom gave the command to Dark to come with him to hit the first tiger shark before it got too close and thus distract those following into attacking their fellow. Jack was left behind in command at the channel entrance where the second birth was now almost completed.

The dark-striped tiger shark that advanced toward Tom and Dark was about twelve feet long, a creature of grace and beauty, but armed with deadly rows of sharp teeth, one gash from which would rip open a dolphin's belly. Tom approached it fearlessly but cautiously and Dark watched his father's technique, holding back a little higher in the water to be ready for the second and hopefully fatal blow.

Suddenly all four hundred and fifty pounds of big Tom's

muscular body dove from his higher position in the water, always an advantage the dolphin has over a shark, who can only move well on one level because of his vertical tail fin. The rapidly beating horizontal flukes of Tom's twisting body drove him down like a projectile till his hard bony snout hit the shark on the left side right over the gills. The terrific blow temporarily paralyzed the shark, though he snapped his great sharp teeth together with a loud clap. An instant later Dark hit him a second blow, on the side but not on the gills, though powerful enough to send him spinning and flapping in the water. Tom whistled angrily at his young son for not hitting the gills, dove again and this time hit the shark in the throat, and a cloud of blood burst from his mouth. Tom had barely time to get out of the way when two other sharks fell on the first with incredible fury, biting and slashing, the sea around them turning dark with clouds of blood and torn flesh.

Tom and Dark retreated to the entrance to the channel between the great rocks and there found Jack and Roger just finishing off a large tiger shark that had come too close to the dolphin pack. While many sharks turned aside to attack the one that Tom and Dark had killed, in the process injuring themselves and starting new battles to the death, six other sharks had inexorably followed down the blood scents of the two newborn baby dolphins and were now moving in on the pack.

Tom called two of the females, Lady and Lillian, up to help with the fight, leaving the other females to guard the babies farther up the channel, but he suddenly saw Ralph and Tony, the teenagers, rushing up from the rear channel. Like most youth they were too anxious to be part of the excitement instead of the dull work of standing guard at the rear. Tom had no time to punish them now, but he whistled at them so shrilly and savagely that they instantly turned tail in alarm and rushed back to their posts. Too late! While they were gone a single shark had penetrated the herd from the rear end of the channel and was bent on grabbing Susan's baby, who still smelled of blood. The usually placid Cora, Susan's mother, turned suddenly, striking the shark at the middle just as he opened his mouth to take the baby. Unfortunately she failed to hit him in the gills, which would have temporarily paralyzed him, and he swung around snapping and gashing her side. In the next second a totally shamed Tony hit the shark in its gills, and Martha, who had come up fast too, struck again in the same place, sending the shark away writhing and dying.

All now knew the vital necessity of getting the wounded Cora to a safe place where she would not attract more sharks. Tom's sharp command caused Martha to hurry the wounded and bleeding Cora up a narrow channel to the edge of high water where she could rest until the bleeding stopped. Fortunately the tide had just changed and was surging in toward the shore and the blood was washed up onto the beach instead of being sucked down into the sea. Martha was lapping it up almost as fast as it flowed, her saliva acting to coagulate the wound.

Meanwhile, in the narrow entrance to the channel, the porpoises were facing only about three or four sharks at a time, and Tom, Jack, Dark and Roger, with Lady and Lillian acting as back-ups, were keeping them firmly at bay, without allowing themselves to get wounded. At last, frightened by the fierce attacks of the dolphins, who had been driven to frenzy by the wounding of Cora and the danger to the babies, the sharks began to drift off into the deeper blue, though some were still tearing at their own wounded and dying in a vast cloud of churning blood.

Peace descended on the dolphins, as they at last lay resting and blowing for air at the surface, while the two new babies were eagerly drinking the milk of the two new mothers. The milk was squirted by muscles that surround the nipples, vital for babies that are fed underwater between breaths and that have little time for sucking. Both babies were blissfully unaware of the recent near approach of death. But Tom, the huge herd bull, was slapping two terrified young dolphins, Ralph and Tony, with his hard hindflukes, emphasizing with each blow, as they cowered against a rock, the need to obey commands hereafter!

Indeed the life of the pack or herd depended vitally on its inner discipline and social cohesion, as was shown two months later, when Tom's signals, on the coast just south of San Francisco, told his people they had been detected by a pack of killer whales. He had heard, before his comrades, the deeper clicking made by the killer pack as they searched the water for prey. Killer whales sometimes influenced the bottle-nosed dolphins to stay fairly close to shore when they were hunting for fish, as the far larger killers were afraid of shallow waters and rocks. To the great rocks that stood out from shore on the San Mateo County coast, Tom led his herd at high speed. They had scarcely come into the shelter of the first rock, when the towering five-foot-high fin of a big bull killer whale ripped through the water surface at about thirty miles an hour. Against such a creature Tom knew no dolphin had a chance. The blows from the hard snout that paralyzed even a large shark would be useless.

But Tom had wisely led his herd into the shallow waters near the big rocks and was happy to see the huge black fins of the killers sheer off to right and left, as they sensed with their sonar the shallows ahead. The dolphins watched complacently from five- to six-foot-deep water where a killer whale would quickly be stranded and helpless in the waves.

Characteristic of the dolphin herd was the playfulness of nearly all the individuals, a sign of high intelligence in mammals. In September, when the weather was right, and they had comfortably digested a bellyful of fish apiece, enough to make them playful but not lethargic, and when their sonar told them that no danger was near, they selected a sheltered place along the coast where man was seldom seen, and began to play.

In these times of play the characteristics of each dolphin often became evident. Tom, and Jack, the bigger son, would have also liked to play, but they usually felt it was their duty to stand guard, watching the play, but keeping their cones of sonar clickings beaming toward the sea. Aloof and alone at either end of the play area, their flukes moving slowly to keep them in one spot against the waves and currents, they stood guard for the safety of all.

The six youngsters of the herd, all under two years old, played in a special area near the shore below some cliffs where they could best be guarded and watched by their nurse, Julie, whose own baby was among them. All of them, even the youngest, were good swimmers by this time, and Julie had to keep a sharp eye on them to make sure none strayed. At the first sign of naughtiness, her shrill whistles not only warned the children, but alerted other mothers to turn toward her and drive back any culprits to the designated playground. The older youngsters, like Tim and Katchy, found this rather frustrating, but always headed back when warned, and resumed their usual games of tag, hide-and-go-seek or "play with the stick or feather." This latter game worked best with a feather stolen from some seabird, because of its lightness and because it was fun to sneak up under the unsuspecting bird, as it landed on the water, and grab a feather. But a stick did well enough, and one of the young dolphins would seize it from another and swim off at high speed, to be chased by all the others until one of them captured it from him or her. Then the chase would start all over again. A variation was for two of them to play catch with it, tossing it back and forth above the waves, while the other leaped to seize it in between, their bodies flashing in and out of the water like bouncing balls. Tim, the two-year-old, was the supreme starter of new games, such as scaring crabs by suddenly sweeping down on them, directing a strong current of water at them with a quick move of the tail flukes, then swirling about to see if they ran or lifted their big pincers and acted aggressively. Maybe the game meant that the more crabs you made run the more points you got!

The teen-agers, those from about three to six, played courting and fighting games, and copied their elders in other ways, half-serious and half-playful. The young males, of course, had a great time scaring the young females, though one often interfered with another's chase from either jealousy or a desire to tease, and this sometimes started fights in which little harm was done, simply because they did not know how to be nasty to each other yet. But being butted by a hard nose could be painful! Ralph was shy with the girls and Tony was brash and overconfident. In fact Tony once found himself suddenly slammed against a rock by an expert body block from old Tom, who decided he was getting just too obnoxious. Tony was already covered with the marks of such bruises and a few bites as well, but he never seemed to learn.

The adults mainly did porpoising at first, which means forming a line and moving like one body in a kind of dance, each couple or even three or four flashing up out of the water into an arc, down into a shallow dive, and up again. Sometimes they seemed to be trying how high they could jump, some pairs leaping like one, ten feet or even fifteen feet high out of the water. Then they did follow-the-leader, trying various tricks such as moving sideways, or jumping straight up into the air and coming down tail first. Cora discovered a cave that led underwater beneath a great rock and she led them through it time and again at breakneck speed.

All this was part of a glorious afternoon, but there were two who had more serious though equally delightful business. Dark and Darcy, a young female from another herd who had attached herself to this one, had gone off to an area where they could not be seen, to do some serious courting. At first their courtship up and through and out of the green blue sea looked like some kind of fight, for they would crash furiously into each other, chase and nip, turn and nip back, butt heads and noses, strike blows with their flukes and really look as though they disliked each other intensely. Round and round their gray blue bodies swirled and up in tremendous jumps, as if one were trying desperately to escape the other.

But gradually the roughness changed to the more gentle dance of real courtship, the two swimming together in tandem down through the beautiful blue green waters and through the waving feather fronds of the seaweeds, then suddenly increasing speed to hurtle up and out of the surface and twenty feet through the air in perfect harmony. Soon Dark turned his body to swim upside down just below Darcy, the two finally touching bodies, skin to skin, then gently nosing each other, the ardor of love growing until the mating act was reached, over in a few seconds with the swiftness of all the whale folk, though repeated frequently.

In this swiftness of mating, the frequent butting of heads and noses, the herd structure with a dominant bull, and many other features of the physiology and physical structure of the whales (such as several stomachs for digestion) they so resemble the land herbivores, particularly cattle, that some scientists believe the two orders of mammals may long ago have had common ancestors. Many whales carry within their bodies remnant bones of what must have once been hind legs so it seems fairly certain they once were land animals and gradually changed into sea mammals and from plant eaters became hunters of fish and other sea life.

In a way similar to that of wild cattle, Tom's herd of bottle-nosed dolphins would occasionally meet other herds of the same species in the California coastal waters, and the two herds merged for a few amicable hours, playing and getting acquainted, then parting as the dominant bulls grew edgy over the possibility of losing some of their followers. And of course, this did happen frequently, the change-over unobtrusively done and resulting in some good, the mixing of bloodlines and the prevention of too much intermating within one family.

On other days Tom's herd found equal fun and adventure in rushing through the sea to where a large launch or small yacht ploughed merrily over the waves, and rising up out of the depths to the sudden surprise of the people on board and riding the bow waves, something other kinds of dolphins do even more frequently. Each dolphin rode his wave as a surfer rides the giant combers on a beach, letting the wave carry him. The fine act of balancing appears very easy but actually requires that the dolphin's body take just the right angle to the wave and use its power to be carried along it.

We can leave Tom and his companions riding the bow waves of a ship in the Pacific Ocean, sure that we have seen only a small beginning glimpse of the lives of very intelligent and interesting creatures, knowing there are many mysteries yet unsolved, and that man has still to learn how to get into harmony with and learn wisdom from these unique and marvelous beings!

Northern Right-Whale Dolphin

[Lissodelphis borealis]

Description   Up to about eight feet in length; lacks a dorsal fin, which is probably why it is named after the right whale, though completely unrelated. This makes it distinct from all other small-toothed whales along our coast. The lower jaw sticks out beyond the upper jaw. Color black above and glossy white below, including white on undersides of dark flippers and flukes.

Range and habitat   A rather rare deep-sea dolphin, found from southern California north to Alaska.

Habits   Probably hunts fish and squid.

Pacific White-sided Dolphin

[Lagenorhynchus obliquidens]

Description   Length about seven to eight feet, but may reach ten. Beak only about two inches long, not easily seen. Strong back and belly ridges near the tail end of body; bluish to greenish gray to black on upper parts, except for white stripe on rear end of strongly recurved and large dorsal fin; broadly striped grayish white on sides, with belly whitish. Each jaw has twenty-two to forty-five pairs of teeth.

Range and habitat   A common deep-sea West Coast species from Panama to Alaska, usually in fairly large schools that migrate with the fish they chase.

Habits   Has proved easy to tame. Schools often swim in beautifully synchronized ways, the pairs diving and jumping in unison. They like to ride the bow waves of ships.

Grampus, or Risso's Dolphin

[Grampus griseus]

Description Length about eleven to thirteen feet; no beak, but head bulbous and bulging in front. Front flippers rather sharply pointed; the dorsal fin black and strongly recurved. No teeth in upper jaw, but two to seven pairs of teeth in lower jaw. The black flippers, flukes and dorsal fins contrast sharply with the grayish body; more whitish below and with whitish head and throat. Body often marked with easily visible white scars, probably caused by fight with other grampus. Looks like pilot whale, but far lighter in color and with a less bulgy forehead.

Range and habitat Rather rare denizen of deep sea; often found alone or in small groups of only a half dozen or less.

Habits Feeds mainly on cuttlefish, possibly at considerable depths.

Grampus, or Risso's Dolphin [*Grampus griseus*]

Killer Whale

[Orcinus orca]

Description   Males up to thirty-one feet in length with immense triangular-shaped dorsal fins three to six feet high; weight two to four tons; females up to twenty-one feet long, weight 1,500-2,500 pounds, dorsal fins up to two feet tall, curved backward on front edge. The colors are most striking, black or blackish above, white to yellowish below, but also with white pattern extending up side and a large white oval spot above eye. A grayish area seen behind the dorsal fin on the back. The high dorsal fins of the males sometimes have their tops flop over because there is no bone in fin. Snout is blunt with distinct short beak. Ten to fourteen sharp teeth in each jaw, interlocking when brought together by mouth shutting, producing an efficient trap for animal food.

Range and habitats   Found in all of the world's oceans, though rare in the Arctic, and common all along the Pacific Coast, preferring open seaways, including Puget Sound, and avoiding shallow water.

Habits   A very efficient swimming machine with speeds up to and perhaps past thirty miles an hour (thirty-five, says Jacques Cousteau). Hunts food mainly in packs of three to thirty or more, like wolf packs with a leader, even ganging up on large whales and trying to bite their tongues, flippers and other delicate or vulnerable parts, but more likely to attack isolated, sick, wounded or young animals. Main food, however, is seals, squid, dolphins, and other smaller whales. Sometimes catches fish but prefers warm-blooded animals. Extremely few recorded attacks on men. Most food is swallowed whole or in large chunks.

Killer Whale [*Orcinus orca*]

A Life Story

PRELUDE

Despite the fact that killer whales have now been kept in captivity for several years, and that man has met them everywhere on the high seas, astonishingly little is known about their lives. This is partly because they are considered too small for commercial whaling, so less scientific attention has been given them. It is not known whether they migrate or not, though some scientists feel they may swim southward in winter and northward in summer. It is likely that packs, at least in coastal waters, have semi-definite territories. The fact that some killer whales have been found trapped in leads between ice floes or in isolated holes in the ice, as a freeze destroys their chances to escape, demonstrates that at least some of them stay near the edge of the pack ice even in winter. Actually the large dorsal fins of killer whales, especially those of the males, make too close association with ice dangerous, and it is possibly their pursuit of such ice-dwelling whales as the beluga and the narwhal that draws them into these areas, sometimes to their own detriment or even death.

Killer whales have many varieties of clicks, whistles, and other sound signals for both echo-location of prey and for communication. In captivity they have proved highly intelligent in learning tricks, solving puzzles and co-operating with man, probably with even greater promise than the bottle-nosed dolphins. Thus there is a good likelihood, though still far from proved, that they may be a species of intelligent mammals with some kind of real language. Many observers have claimed that they signal each other how to work together when they are attacking other sea creatures and that they have actual planned and organized methods of attack. I myself witnessed what appeared to be an extremely efficient attack on sea lions in the Pacific Ocean off the Farallon Islands near San Francisco, when several sea lions were caught and eaten.

They were once believed to be as dangerous to human beings as sharks, but more recent evidence has shown them to be unlikely to attack people, even those swimming alone in the water near them, who could easily be attacked and killed. There are a few records of their attacking boats, but usually when they are attacked by man first. In one such case near Bodega Bay, California, the boat was completely wrecked, but the crew was able to swim ashore unharmed. Perhaps they either can't quite figure us out or actually feel some kinship with us.

There are records of wild killer whales genuinely cooperating with men in attacks on larger whales. For example, on the coast of New South Wales in Australia, at the town of Eden on Twofold Bay, whale hunters for many decades, up to about 1930, were helped by killer whales to capture and kill humpbacked and right whales. The killers would drive the large whales into the bay where the men could kill them more easily, and the men reciprocated by letting the killers tear off the tongues and lips of the large whales that had been harpooned and killed. One huge killer bull, called Old Tom, helped with this job for over forty years and became quite famous!

ADVENTURES OF A KILLER WHALE PACK

The Kwakiutl Indians of the British Columbia coast were famous fishermen and deep-sea hunters and the animals of the sea they most admired were those other great hunters, the killer whales. They called the chief of all the killer whales Kago Laye and often put his figure on their totem poles.

In this story we shall call the chief bull of the killer whale pack Kago for short. It was a good-sized pack of about twenty-two killer whales, including the three sons of Kago, who were all smaller than and subservient to him, six females that were his harem, one cow for each of the sons, four adolescent whales, and five babies. But Kago, over four tons in weight, thirty-one feet long and with a six-foot-high black dorsal fin that was his badge of chieftainship, ruled the whole pack with every turn of his massive head and the clicks and whistles he sent through the waters. It was the great triangular back fin that sent shivers through the heart of any lone bull killer whale who might have thoughts of challenging his rule. Such whales usually wandered on when they saw that great fin cleaving the dark waters, the many lesser fins of his pack close behind. But if one of them had sensed that Kago was wounded or ill they might well have challenged him.

The territory patrolled by Kago and his pack was a long one of many miles along the coast where the Kwakiutl used to rule, in what is now British Columbia, from Cape Cook and Quatsina Sound in northwestern Vancouver Island, north to Smith Sound and Cape Caution on the mainland, and east into Queen Charlotte Strait and Sointula Island. It was an area of many deep inlets lined by rocky cliffs and mountains like the famous fiords of Norway, and far inland could be seen, even in summer, the eternal snows of Silverthrone and Monarch Mountains and, highest of all, Mount Waddington. For nearly twenty years Kago had ruled the blue green waters of all this area, and he and his pack knew all the hidden bays and inlets, the deep places and the shallows, the rocks and beaches favored by seals and sea lions when they hauled out to escape the killers, the places where dolphins liked to come, and, far out to sea, the pathways of the migrating large whales and the fur seals, all places where there was good hunting for his people. The Kwakiutl of long ago knew that the killer whales were an intelligent people of the sea with a secret place deep in the waters, where legend said they dressed like land people and talked together about the rule of the ocean waves. If one could go there and make friends with them, it was said, they could give you magic powers to become a mighty hunter or medicine man, but if you failed, terrible would be your fate!

One day Kago led his pack northward farther than they had ever been before, up the straits that lead by Namu and where the famous big ferries of the inland sea come down from Alaska, or go northward with hundreds of tourists lining their rails. One of the ships was passing northward, paralleling the swimming pack, and at least two dozen binoculars were trained eagerly over the ship's side on the black fins sliding through the water, the ripples pouring away from each fin like long lines of light in the glittering waters.

It was the mating season for killer whales, May to July, and also the time when babies were born. Two of Kago's wives had given birth a few days before, and now he was calling to the first that had come into heat after the birth. It was a peculiar deep whining call that resounded through the waters, a low-key up-and-down singing, that was a command from the chief. He never moved toward the female. Instead he swam away from the pack, almost out of sight of them. His wife, the third in rank of the six, was hesitant for a while, being a little coy, but the insistent calling got to her and finally she came to him, sweet and obedient. Then they began the ritual of the mating dance, the building up of passion by swimming over and under each other, touching and rubbing bodies and flippers and nose, the great bulk of the herd bull at least twice that of his consort and over three times her weight. Even near him she was still shy and reluctant at first, fleeing when he came too close, protesting with shrill little cries, but he kept calling and calling until she came closer and they touched again and again, at last to merge side by side as one being.

When the mating was over the female went quietly back

to her baby, which another had been taking care of while she was gone, a sort of nurse. Kago began to send out low-pitched sonar clicks into the sea. These were a sign of hunger and a sign to the other killers to close ranks and respond with cones of sonar sounds, searching the inland waters for prey. The clicks were low-pitched at first, for deeper sounds travel much farther than the high-pitched clicks used for close-in quality sensing.

In a few minutes the clicks told them that large bodies were moving through the waters to northward, and Kago signalled the pack to spread out in two lines to left and right and about two hundred yards apart. They moved at near top speed, or better than thirty miles an hour, while behind was left the little group of three females and five youngsters, the latter too small as yet to take part in a fast chase.

Whoever was moving ahead of them was doing about fifteen miles an hour and heading for a rocky beach. As the sonar clicks told Kago and his pack that they were getting nearer their prey, the frequency changed from low to high and the high-frequency clicks began to tell them qualities such as size, shape and texture, so they soon knew they were chasing eight small seals of around 150 to 200 pounds in weight, probably harbor seals. The race became a test of whether the seals would make the beach before the killers reached them. It was also a weeding out of the weak from the strong, which is how harbor seals maintain a healthy, active population, since only the fittest had a chance to escape the killer whales. This is different from the shooting of seals by men, who kill the strongest and most healthy as easily as the weakest, though perhaps not always the smartest.

The eyes of killer whales can adjust both to under the waves and above by using eye muscles that change the size and shape of the eyeballs to fit the needs of the two different media. So the killers porpoised now and then out of the waves to look ahead and see the streaks in the water that marked the fleeing seals. As he saw them, Kago pushed up his own speed and called on his pack to increase theirs, his own vast form literally splitting the waves as he pushed into the lead and closed on one of the racing gray bodies.

Whether he was actually guiding his pack with a language somewhat like ours we don't know yet, but even if it was simply the emotion in his whistles and the speeding up of his clicks, or the way his body was acting, his signals drew the two lines of killers together around the fleeing seals. Soon he was close to one and opened his vast mouth with its long white teeth. The seal dodged, but too soon, giving Kago a chance to swerve with it, with incredible dexterity for so huge a body. The next instant his teeth snapped together, bringing instant crushing death. The seal was tossed into the air, a limp and bloody morsel, then swallowed whole as it fell.

One of Kago's sons was having a harder time. His seal was a more cagey beast, wise in the ways of killer whales. She dodged at the last possible moment, just as the big-toothed mouth was opening behind her, and the teeth snapped together, tooth grooved to tooth, a perfect trap, but with no seal in it! The furious killer swirled after her, opened his mouth and lunged again, snapping with a clang like that of a steel trap, and again he missed! A third time he tried, and would have flung himself in a rage up on the beach rocks she had just reached, if his mighty father had not somehow sent him a warning that stopped him before it was too late.

Meanwhile Kago had caught and killed another seal, whose remains he left for the rest of the pack to feed on, and three others had also been killed in the chase, although three had escaped. As the last meat was chopped into huge chunks and swallowed by the pack, a new sound in the water made them turn almost as one and face outward from the beach. Catching the echo-locating clicks of a different pack of killer whales approaching, they moved slowly out to meet them, needing deeper water in which to meet trouble.

In the deep channel the two packs cautiously approached each other. Kago knew he was outside his territory, but he sensed from the echo clicks that his pack was the larger, and his pride made him doubt there would be another bull his equal. However, he also knew that the other pack had the psychological advantage of being on their own turf, and that a clash between packs, with the inevitable death of several of his own followers, did not make good sense unless the issue was vital. So he moved out in front of his pack, all the rest watching and listening to him tensely, and turned sideways toward the approaching pack, showing his immense six-foot-high dorsal fin, like the great black flag of a sea kingdom. All watched as the other pack leader came up to show his fin, which was at least half a foot shorter than Kago's. But the way he held it and the way he rested in the water told Kago that if he made a wrong move there would be war.

Kago, without any direct reply to the challenge, gave the signal to his own people to turn south and follow him as he moved in that direction. He moved slowly, leisurely to show that he had no fear, but he moved in the right direction to prevent a fight to the death. So peace between the killers came to Fitzhugh Sound where the town of Namu lies.

I have seen two packs of killer whales facing each other like this in the Pacific Ocean near the Farallon Islands, with the two big bulls out front, their back fins high out of the water. It looked like negotiations about a territorial boundary, but whether this is really the case will probably take much more research to prove than has so far been done.

As Kago's pack turned south during the summer months and moved again through familiar territory, they hunted mainly seals, squids and some of the larger fish, as well as occasional porpoises and dolphins. The striped porpoise, eight feet long and medium fast, they occasionally caught, but the smaller Dall porpoise was so speedy and agile in dodging that it led the killers too merry a chase, leaving them with a splash of the flukes that could be interpreted as derisive laughter!

Far out in the open sea northwest of Vancouver Island, Kago once saw a chance to corner a smaller than usual school of Pacific white-sided dolphins. These swift coursers of the North Pacific are among the most beautiful of all the whale folk, because of the graceful lines of their bodies and fins, even the tail flukes having an artistic, slim-pointed appearance.

Knowing their speed but also their timidity, Kago used the utmost caution in devising a method to corner and then attack them. First he sent three of his people in a direction away from the dolphins, making loud clicking noises to sound as though the killers were moving off. Then he told the pack to use no noise at all, but to guide themselves by sight only into a great circle completely surrounding the dolphin school. The silence was necessary because killer whale clicks are always a signal to dolphins that their most deadly enemies are approaching. Forming the circle and gradually closing in on the school would confuse the dolphins when their sonar picked up the body forms of the killers on all sides, and they would not know which way to turn.

The pack spread out to form the circle as directed, the three who had gone away clicking soon returning to join them in silence, and the attack went according to plan. Soon the killer whales were surrounding a dolphin school of about twenty-six individuals, including mothers with their young. The dolphins saw the giant black-and-white forms on all sides, grew frightened and confused, darted now in one direction and then another, but always came back to the center of the circle. Against sharks they would have attacked with snout blows to the gills, but against the killer whales they felt helpless, their will to resist paralyzed by a terrible fear.

Suddenly one of Kago's giant sons dashed forward and, leaping into the midst of the dolphins, seized one and almost sheared it in two with one crunch of his powerful jaws and teeth. Shrill whistles of terror burst from the dolphins and they shot off in all directions to escape, but were seized and killed or even swallowed whole by the surrounding killers, Kago swallowing one and cutting another in two in two quick onslaughts. Of the twenty-six, perhaps eleven escaped by fleeing at full speed through the sea, none of the terrified creatures making any effort to resist the attack.

But the killers' trip out into the deep sea did not leave them unscathed. The next day a large pleasure launch with two powerful engines came toward them from Vancouver Island, chasing them to see how fast they could swim, the prow of the boat sending two high white lines of waves V-ing across the sea. They almost pulled away from the boat, but not quite, and as they tired, there were suddenly two men in the bow of the launch with high-powered rifles.

Kago felt a sting of sharp pain on his back as a bullet grazed it, and he gave the signal to scatter, dive deep and escape. He himself dove down only twenty feet when he heard the whistle for help from one of his wives. She was floating in pain near the surface with two bullet wounds in her side. He no sooner came up beneath her to support her when another bullet struck him, this time through the lip. He turned his head toward the ship in sudden anger, about to call the pack to attack these strange enemies, when he caught sight of a smaller figure beside the two men, who suddenly pushed their guns up into the air just as one fired, and the bullet from his rifle bounced on the waves, hitting nothing. Sounds of fierce argument followed, but the one who had stopped the shooting apparently won, as no more shots were fired, and the people on board watched Kago and one of his sons supporting and comforting the wounded female. Blood was flowing from her rapidly, and she gave a sudden last gasp and died.

"You see!" cried a woman from the boat. "He was brave! He came to help his mate even when he knew you might kill him too. Let him live!"

In the sea Kago, with a sore back and a bleeding lip, swam slowly and sadly away. He knew the sharks would gather soon.

The killer whales are hunters and killers of other sea mammals, but they have a function in the sea they have had for long ages before men intruded on their world: to keep the other whale folk in balance, not to let them increase too greatly in numbers and to kill off the sick and wounded or those too weak for the wildness of sea life. The healthy agile stock of each species is thus preserved. For untold centuries the larger whales had no enemies except the killer whales and possibly a few of the larger sharks, to keep them from overpopulating the seas, which might lead to starvation and sickness. The killer whales, like men and the other more highly developed and social hunting animals, such as wolves and wild dogs, have to use a fine degree of intelligence to outwit their prey, especially the large or swift whales. This has helped evolve them into one of the most interesting and highly organized of all animals.

In the fall Kago and his pack moved out into the deep sea again to take their toll of the different whales that migrate down the Pacific Coast from the great feeding grounds in the Bering and Chukchi Seas, where the krill dance and feed in their billions under the sea. Sperm whales the pack left strictly alone, except for a rare wounded or sick individual. They knew these giants had long sharp teeth like their own, and highly cooperative herds in which the great bulls protected the females and young most effectively. The largest of all animals, the blue whales, eighty to one hundred feet long, and the fin backs, sixty to eighty feet in length, were generally too large to be attacked except any sick or wounded by man, or if a young whale could be found alone, far from its mother. Most of the other whales were fair game for a pack as big as Kago's, and with such a sagacious and brave leader; though each had to be treated with respect as potentially very dangerous and let alone altogether if too many were together.

A dark brownish sei whale, distinctively dark colored all over, and fifty-five feet long, was spotted by the pack one late November day, heading south alone. They closed in, feeling they could handle her, but learned sei whales have greater speed than any other of the large whales. Clicking furiously the pack set out after her, but her long streamlined body, marked by its unusually high dorsal fin for a rorqual, seemed to gather power like a motorboat switching on a second engine, the small but well-balanced and perfectly formed flukes moving with the rapidity of a propeller. Up to twenty-five, thirty, thirty-three miles an hour, they flashed through the water after her, wild with the thrill of the chase, but amazed at how easily she drew away from them! Like wolves outrun by a powerful buck deer, they finally gave up the chase and rested, blowing for breath at the surface of the sea, and searching around with their sonar for an easier prey.

They soon found what they were looking for—in a school of the smallest of all the rorquals, the sharp-headed finner whales, also called little piked whales or minke whales, averaging about thirty feet in length, and none of them bigger than Kago. These little whales, marked by their sharp-pointed heads and a large white patch on each forefin, their bodies black above and white below, were swimming south to warmer waters. But with thirty or more of them in the school, they were by no means easy prey. They were fast swimmers and had tough skins and good courage, for they had traded blows with killer whales for millions of years, and each species knew the strengths and weaknesses of the other.

Hearing the killers closing in on them, the school of sharp-headed finners upped their speed and closed ranks, the babies in the middle borne along partly by the water pull of their mothers' bodies, the bulls on the outside. They had no teeth, but they were quick with blows from their tail flukes, backed by four tons or more of weight, and not to be taken lightly. They knew they could not escape by greater speed like the sei whales, nor could they get away by climbing out on land like the seals. They knew they had to flee but also be prepared to fight.

With marvelous swiftness their tail flukes beat the water as they fled to the south, Kago and his pack in full cry behind them, and gradually overtaking them like wolves closing in on a herd of caribou. The almost equal speed of the two kinds of creatures made attack difficult, for the finner whales could not only dodge at full speed but sometimes execute a body block or lash out with their flukes. It was the bulls of the killer whales who led the attack when they got close, and the bulls in the fleeing herd that defended against the attack. Again and again the killers darted in at the moving mass of finners, trying to seize a fin or get a neck grip, and again and again they fell back frustrated, as a finner swerved his tough body and recoiled from a bite or dodged at the last possible second, leaving the killer gnashing his teeth on empty water.

It was Kago who finally drove in close to the front of a finner bull and suddenly dashed in with enough speed to seize a front fin in his powerful teeth, jerking him away from the swimming herd. Even then the finner might have escaped, for he jerked loose again with a powerful heave and his flukes struck out and nearly broke another killer's jaw, but the others all ganged up on him at once, biting and tearing, and he died with two gushes of blood from blowhole and mouth.

Leaving this meat to the lesser members of the pack, the killer bulls took up the chase once more, Kago in the lead, and again closed in on the finner herd. This time it was a female who dropped back from the herd, too tired, perhaps from sickness, to keep up the pace. The herd had tried to help her, but it was useless, and she had to be sacrificed to save the lives of the rest, particularly the young ones. Such is the cruel but necessary law of the sea. As the killers encircled her, she struck one a blow on the head with her flukes that stunned him. His mate rushed up and pushed him to the surface and held him there for air, or he would have drowned. But meanwhile the female finner had died that her herd might live. By the time the killers had torn her to pieces, the rest of the herd was too far away to be chased anymore.

Resting in the waters, Kago sent back low-frequency sounds to the females and young of the pack, who were far behind, telling them to come up for a feed, and he and the other larger bulls formed a ring to guard the meat that remained from any attack by sharks. The sight of that fierce ring would have been enough to send any shark scurrying away from a danger too great even for its mad hunger. So that day the law of the sea was obeyed, the law of the herd (the finner whales) and the law of the pack.

Sometimes Kago sent out scouts to look for food, each directing the cones of his or her sonar to seek echoes down through the depths of the sea or out at all angles. Twice they found little pods of humpbacked whales, once two families and a nurse whale together, another time three families, but these tough-skinned fifty-footers were too quick with the power of their flukes and their immense clublike front flippers for Kago to risk the lives of his pack against such healthy and alert combinations. But when the scout sent back word, by the deep clicks that carry far through the waters, of a lone gray whale moving slowly south, the pack closed together and moved toward the kill. The gray was an old bull, wounded and sick, doomed soon to die by sharks if the killers had not found him.

When he heard the sound of their approach, fear overcame him so completely that he lay paralyzed in the water with no attempt to flee, for he knew flight was useless at his slow speed. With hungry eagerness they rushed upon him, tearing first at his tongue and lips as he lay still in the waters. Soon he was moaning, with the strange sad moaning of a hurt whale; so let us turn away from his death, for it was not a pretty thing to watch or hear. It was however much better than the long slow death of pain and suffering that would have been his if no "sea wolves" had found him.

Possibly this is one of the reasons for the occasional strange strandings of lone whales on the beaches. Whole herds of whales, or at least parts of herds, are known to get stranded on beaches, but this is probably caused by mass hysteria in storms, or in the strange panic that overcomes them when their sonar gets fouled up on the sloping beaches of sand or mud and they cannot use their echo-location to detect the source of danger. But as we have seen, lone whales who strand themselves voluntarily may possibly be suicides, preferring to die by desiccation on a beach than be torn to pieces by hungry sharks or killer whales.

Kago himself one day came near to being food for sharks when a huge strange bull came up out of the sea and challenged his leadership of the pack. It was a bull who, when younger, had been driven away by a pack bull and wandered the sea alone for a time, gathering strength and courage to fight for a pack of his own, or, as some did, simply steal a single cow. But this bull did not sneak around the fringes of the pack, trying to lure a female. Instead he flaunted his five-and-half-foot-high dorsal fin before the whole pack, until Kago brought his great teeth together in smashing unison like the steel teeth of a giant digging machine clanging together when they meet in rock and soil. The shock of the sound waves from that clashing echoed for a mile or more through the sea, but was echoed by a savage smash of the teeth of the strange bull. So the two monster animals moved warily toward one another, gradually narrowing the circle of movement, head to head and eye to eye, trying to outstare each other. And suddenly the stranger struck with a great lunge of his body and a fierce snap of his teeth. He was younger and quicker than Kago and just as strong, and, if he had at once seized the front flipper and torn it badly, Kago's days would have been counted in very small numbers.

But Kago was a veteran of a score of such battles, and he reacted by moving sideways just enough to be missed, then closed mouth to mouth with the other bull, tearing and shaking. Like two great bulldogs locked together, they reared high on their lashing tails and each tried to jerk out a chunk of flesh. Kago's extra weight and size at last made the other bull give way, part of his jaw torn off, and he turned and fled through bloody foam into the distant sea. Kago shook his head groggily, for he also had been wounded and his blood was on the waters. But his wives clustered around him, and his pack was still with him to help him heal. Perhaps Kago felt his age at last and knew there would come a time when he would not win.

False Killer Whale

[Pseudorca crassidens]

Description Length of male nine to nineteen feet, of female seven to sixteen feet, with the smaller sizes more common in the north Pacific. Distinctive all-black color,

False Killer Whale [*Pseudorca crassidens*]

though rarely with a few star-shaped white spots; but head sloped backward instead of bulging like similar all-black Pacific blackfish or pilot whale; the recurved back fin near middle of body. Eight to twelve pairs of prominent round sharp teeth in each jaw, not oval as in killer whale. Light scars may be present, indicators of fighting, probably between males.

Range and habitat   Found in all open seas except Arctic, in herds or packs usually of a hundred or more, sometimes numbering thousands.

Habits   Feed mainly on large fish, such as tuna, bonita, mackerel and mahimahi, which are fast swimmers, so speed of false killer whale must be considerable. They also feed on squid. Schools may follow a leader whale, probably a large male, which may account for occasional mass strandings of this species on shelving sand and mud beaches where their echo-location does not work. They are found to adapt very readily to captivity and get along well with other delphinids as well as learning tricks quickly. The species has a well-developed method of communication, with not only echo-location clicks, but many kinds of squeaks, whistles and other short cries.

Pacific Pilot Whale, or Pacific Blackfish

[Globicephala scammoni]

Description   Length of male up to twenty-two feet, female up to sixteen feet. Prominent bulging forehead is distinctive, as is nearly all-black color, mixed only with a narrow white stripe on the belly, and sometimes a gray saddlelike area just behind the back fin, which is long, low and strongly recurved. This fin is closer to the head than to the tail, which is quite unusual in toothed whales (except for killer whales). Flippers relatively long and slender; a very short beak. Eight to thirteen pairs of comparatively small teeth found in each jaw, each tooth around a half inch in diameter. Males often scarred by fights.

Range and habitat   Found in most of north Pacific and on our coast from Alaska to northern Baja California, usually

Pacific Pilot Whale, or Pacific Blackfish [*Globicephala scammoni*]

well offshore and generally in schools of six to fifty individuals, but occasionally in much larger schools of hundreds.

Habits   There is some indication that they may migrate north or south with changing water temperatures. Calves at birth are around five to six feet long and are born mainly in the summertime. Mating is likely to occur in spring; gestation period is around fifteen to sixteen months; calves are generally weaned after about twenty-one to twenty-two months, which shows an unusually long period of caring and training by the mother, generally an indication of a highly intelligent animal. However, pilot whales are noted for their stranding on beaches in fair numbers, especially on beaches with sloping approaches where echo-location does not work well. This is believed to happen because each school is dominated and led by a large male, whom the rest of the school or herd follow blindly. Females become sexually mature at about twelve feet in length, males at around fifteen or sixteen feet. Maximum age is around forty to fifty years.

The species seems to be polygamous, with females outnumbering males by as much as three to one, and the largest and most powerful males dominating harems. Fights between males for dominance often take the form of butting each other with their bulging foreheads like bulls do but also by biting. Squid and cod appear to be major food items, but other fish are undoubtedly captured. When a pilot whale blows, it may produce a kind of bark. They probably have many other methods of communication, as recent observations of them in captivity show signs of their quick intelligence, and they can learn tricks just as well as the bottle-nosed dolphin.

Harbor, or Common, Porpoise

[Phocaena phocaena]

Description   Length four to six feet, weight up to 110 to 160 pounds, being the smallest of our whales. Has a distinctively stout body, with no beak, and a low and triangular-shaped dorsal fin; blunt nose. Color blackish or brown above, except for light-colored edge to back fin; white on belly and whitish behind eye; a weak dark stripe reaches from corner of mouth to front of flipper. Has twenty-two to twenty-seven pairs of spadelike teeth in both upper and lower jaws.

TOP: Harbor, or Common, Porpoise [*Phocaena phocaena*]; BOTTOM: Dall's Porpoise [*Phocenoides dalli*]

Range and habitat   Found near coast and in harbors and bays from Alaska to Pismo Beach, California, and perhaps farther south, liking shallow water. May swim up large and even small rivers.

Habits   Rather sluggish swimmer, probably escaping killer whales by getting into water too shallow for them to follow. Found mainly in small schools of five to fifteen individuals, rarely larger ones. Feeds on sole, herring, whiting, and other sluggish bottom fish, also on crustaceans and plants. Lives to about thirty years.

Dall's Porpoise

[Phocoenoides dalli]

Description   Up to six or seven feet long, with rather stout body and blunt nose; weight about 180-220 pounds; back fin triangular, but higher than in harbor porpoise; visible bump on lower back. Color generally grayish black above, though often greenish tinge on jet-black flippers; flukes about one and a half feet wide; dorsal fin and tail flukes with white hind edges, and large white area on middle side of body and on belly. Twenty-three to twenty-seven pairs of small conical teeth in each jaw.

Range and habitat   Avoids shallow waters, ranging along coast from Baja California to Alaska, generally within twenty miles of shore.

Habits   Usually swims in small groups of eight to fifteen individuals, often at amazingly high speed for such a thick-bodied porpoise, up to thirty or more miles an hour. The high speed makes it likely they capture and eat medium-sized fast-swimming fish, such as mackerel, and that they can often escape the attacks of killer whales.

SUBORDER

MYSTICETES

(Baleen, or Whalebone, Whales)

These whales are generally so large that they range from twenty-five to over one hundred feet in length. Unlike the toothed whales who feed on comparatively large creatures, the baleen whales feed largely on tiny invertebrates, though a few capture medium-sized fishes in the process of swallowing the smaller living things. The horny baleen plates or whalebone (described more completely under the description of the humpbacked whale) act as strainers through which the water is expelled from the mouth, leaving the usually tiny living food trapped and swallowed.

The whalebone is made of a tough flexible substance closer to our fingernails in character than to our teeth or bones. The blue and humpbacked whales, which swallow larger crustaceans and even small to medium-sized fish, have more widely spaced strainers in this whalebone than the finback or sei whales, which have filters almost as fine as wool to catch and hold very tiny crustaceans. Most baleen whales take great gulps of this food and then raise their heads above water to let gravity help them swallow it, but the sei whale, and also the right whale, use their large mouths (half-open) and baleen as a skimming tool to collect the tiny crustaceans, such as copepods, that swarm in the waters near the surface. The sei whale can do this at high speed; the right whale is slower. After the skimming the whale dives, shuts its mouth, and swallows the food, forcing the water out through the baleen. The other baleen whales push their heads out of water and turn sideways, so the water is forced out of the side of the mouth before they swallow.

Baleen whales must need a special process to get rid of the salt they swallow in such quantities with the seawater, more so than the toothed whales who live largely on fish and squid, which have already eliminated most of the salt in their bodies. Baleen whale urine contains much more salt than seawater does, so urinating helps get rid of at least some of the excess salt. There may be other ways that we do not know about for getting rid of salt.

Like all whales, baleen whales have milk far richer in fat than do cows—about thirty-three percent compared to a cow's three and five-tenths percent. This is largely because of the vital need for the young whale to build up as much blubber as possible under its skin in order to insulate the body against cold. The blubber also protects against shark or killer whale attacks, and as a reserve food supply.

FAMILY

ESCHRICHTIIDAE:

Gray Whales

Only one species as described below.

Gray Whale

[Eschrichtius glaucus]

Description   Length thirty to fifty feet; weight twenty to forty tons; sexes of similar size. No dorsal fin; two to four short grooves on the throat. General color of adults is black or blackish mottled with gray or whitish patches (which are actually caused by barnacles that cling to the whale's back and fall off in the cold waters of Bering Sea when they die). Young whales may appear slatish black. Mouth appears to divide the head into two halves, unlike other baleen whales, whose lower jaws generally appear much larger than the upper part of the head, and unlike the sperm whale, whose lower jaw is much smaller than the upper head. See page 188 for appearance of spout.

Range and habitat   Migrates yearly between the coastal lagoons of Baja California, also those of southern Sonora and Sinaloa, Mexico (December to March), and the cold waters of the Bering and Chukchi Seas (May to October), where most of the feeding is done.

186

Food and locomotion   Feeds mainly on sea bottom amphipod crustaceans (similiar to beach fleas) in the north, lightly on marine plankton while on migration, and on sardines, crustaceans and bivalve seashells in the Mexican coastal areas. Generally travels at between four to five miles an hour, but can attain a speed of ten miles an hour or possibly more.

A Life Story

PRELUDE

The family of the gray whale is a very ancient one, so that many scientists consider this species almost the equivalent of a living fossil. Its structure, particularly that of the baleen (or whalebone) in the lower jaw, used for screening the food it eats from the mud and water, is much simpler than that in other baleen whales. Also its habit of feeding on shallow sea bottoms, by actually dredging the bottom to catch amphipods found in the mud, would appear to be more primitive than that of other baleen whales who feed mainly in the open ocean on marine plankton. Primitive and fossillike or not, the gray whale appears to be well adapted for its particular way of life, showing considerable intelligence and noted during the intense whaling years of the nineteenth century for its very considerable courage under the brutal attacks of men.

Whalers began to hunt the gray whales intensively early in the nineteenth century, especially in the Arctic seas and during migration, but for some time they were puzzled about what happened to these whales during the midwinter breeding and calf-birthing season. The discovery of their breeding grounds in the quiet desert lagoons of the Baja California coast by Captain C. M. Scammon in 1854 opened up one of the worst periods of wanton slaughter

Gray Whale [*Eschrichtius glaucus*]

in whaling history. Let us look at this period from the standpoint of the whales.

For untold centuries, stretching into millions of years, the gray whales of our Pacific Coast had traveled on their annual migrations between the cold Arctic seas and the warm coast of California and Baja California. During a great part of that time their only serious enemy had been the killer whales, whose packs occasionally attacked them, though not with too much success, except when some of the grays were sick or wounded. This resembles what happens when wolves in a pack attempt to attack a herd of caribou or a group of moose, who usually fight off the attack successfully, unless they are sick, alone or wounded. Sometime during that long period gray whale ancestors discovered the hidden lagoons of the Mexican coast, and found them ideal for the birth and rearing of baby gray whales, because killer whales were afraid to use such shallow waters.

Even when men came to North America, the first, the Indians, left the gray whales alone in their breeding grounds. Only a few of the more daring of the northwest tribes, the Makah of what is now Washington State and the Nootka of Vancouver Island, dared paddle their large cedar-log canoes out into the wild Pacific to attack a few of the grays as they passed by on migration. But the white Americans who came in the nineteenth century were a different breed, men whose overweening greed for money made them track down the luckless grays ruthlessly to their hidden paradise in the desert lagoons, completely oblivious to the intelligence and sensitivity of these whales whose brains are as complicated as those of men.

So one day the gray whales' paradise was invaded by what must have seemed to them literally demons out of Hell. The first ship that found the secret way into Scammon's Lagoon was very soon followed by others and the whales were attacked with sharp and explosive instruments that caused them violent pain. Can you imagine a dumdum bullet exploded in neck, hip or stomach, tearing great holes in tendons, muscles, bones and intestines? No wonder some whales, and particularly the mothers who saw their babies in such danger or so badly injured, turned on the small boats from which these painful weapons came and either threw them over with their heads or stove them in with one mighty blow of their tail flukes. Against these defenses of the whales the men cleverly designed other ways of attack, firing their harpoon guns along narrow stretches of the lagoons through which the whales had to pass but could not reach these two-legged demons. So the years of bloodshed and pain went on, with the whale killers following their prey into every nook and cranny of the many lagoons, until the waters far and wide turned red with blood and the moans of the great whales were greeted everywhere by the excited and bloodthirsty shouts of their human tormenters.

At last, toward the end of the nineteenth century these intelligent animals could stand no more of the slaughter, and were indeed in danger of disappearing from the earth. A signal must have gone out from their leaders that caused them to mysteriously disappear from their former haunts. Then even these ruthless slaughterers decided, with regret only for the money lost, that their victims were now a vanished race. But somewhere, in some hideaway we may never learn about, a few of the gray whales found safety from the fatal harpoon.

In recent times, however, the gray whales seem to have sensed that somehow the worst of their ordeal is over and that an increasing number of their former enemies are becoming kindly disposed toward them. At last conservationists and naturalists have risen to protect them and the first International Whaling Agreement of 1937 brought them safety from the harpoons. And during the 1930s and 1940s the gray whales began slowly to come out from their secret hiding places and follow once more their ancient migratory routes to the desert lagoons they had used and loved so long! Here then is a story of the gray whale today, protected and honored by men, and it is a story told with joy!

STAR LADY AND LITTLE STAR

Early one December some observers in small boats, fifteen miles out from the harbor of San Diego in the Pacific Ocean, part of a flotilla of many gray whale observers, noticed an unusually large gray whale. The experts proclaimed her a female and noticed she was moving faster than usual, probably six miles an hour or more. They thought also that she was pregnant and near full term and so was traveling fast to reach the sheltered lagoons of southern Baja California, where it would be safe to give birth. Some unusually sharp-eyed and powerfully binocu-lared people noticed that she had a starlike blotch on her forehead and promptly named her Star Lady.

Like most gray whales on migration, Star Lady traveled almost continually, never ceasing to swim at night, and resting or sleeping only about six times during each day, and only for about a half-hour interval. The rhythmic movements of her great tail flukes, nearly four feet broad, as compared to her overall length of forty-four feet, kept her body constantly in motion through the water, while her side fins acted more as rudders, steering her in the right direction. Her streamlined body eased her through the water as if she were one with it, and every few minutes, even as she swam, her head and back would surface to allow her to blow out the humid used air from her body, the spout appearing like a small geyser, wider of shape than any other whale's, rising from the sea with an explosive sound and followed by several deep sighs as she inhaled fresh air. The whale's spout is more clearly visible

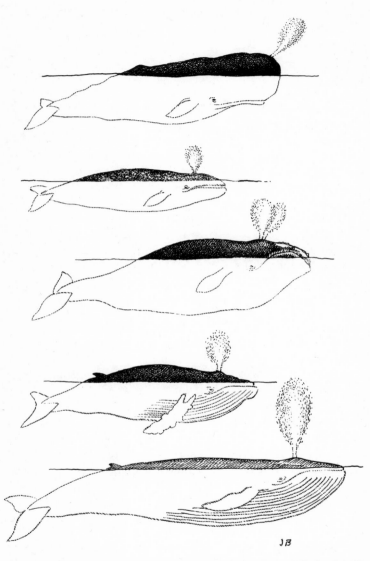

Whales spouting. TOP TO BOTTOM: Sperm Whale; Gray Whale; Pacific Right Whale; Humpbacked Whale; Blue Whale.

than water vapor because of the tiny drops of oil in it, and is visible on hot dry days as well as cold.

Unlike most gray whales heading south, Star Lady made no pauses to sweep in mouthfuls of plankton from the rich California coastal waters, but depended solely on the more-than-foot-thick layer of blubber that covered her body, blubber originally built by feeding on the amphipods from the muddy bottoms of the shallow Bering and Chukchi Seas of the Far North. She traveled south day and night until the clicking sounds of her sonar, bouncing off the shallow depths of the Baja California coast, told her that the secret entrance to Scammon's Lagoon was near. Then she turned inland, and, in the happy-go-lucky way of the gray whales, very different from that of their great kindred, who are afraid of shallow waters, brushed and bumped over the depthless lagoon entrance, often lifted over an obstruction by a wave, until she entered the somewhat deeper waters of the lagoon itself and that maze of waterways that gray whales consider the ultimate paradise for raising babies.

She had scarcely found a good hiding place in a narrow neck of water surrounded by the pale gray lifeless sands of the desert when her body began to jerk and then convulse in the rhythmic contractions of birth. Soon, tail first, as is the natural way for whales, the baby came out, a mere sixteen feet in length and weighing about 1,500 pounds compared to her thirty-five tons. Gently she nosed him to the surface and held him there so he could get his first breath, a great sigh of satisfied motherhood coming within her as she herself rose to breathe. The next job was to nudge and position him, so he could find the hidden nipples that squirted into his soon-eager mouth the rich whale milk.

In the little secret world of water, hidden among the sand dunes where Star Lady's son, Little Star, was born, she nuzzled and touched him every few minutes, as if to make sure he was still there, and so jealous of his safety that even when a great blue heron winged slowly over from the seashore and dropped down on the nearby bank to peer for fish in the pool, she splashed him with water from her flukes to drive him away. As Little Star butted, at first rather clumsily and aimlessly, about the lagoon head, peering with myopic curiosity at the water plants, the sand and the rocks, she would brush away any creature that came near or touched him, and turn him gently away from any sharp rocks. If a whaler had come to attack him, she would, like her female ancestors of a century before, have risen with terrible fury to smash and destroy.

But fortunately the lagoon had the ancient calm of the days before the coming of man, and the little calf grew with extraordinary rapidity, sometimes as much as a hundred pounds in a day. He was putting on an increasing layer of blubber under his skin, to insulate him when in time he would swim north with his mother to the cold Arctic seas off the Alaska coast.

After about three months his mother brought him out of their secret hiding place, to where he could mix with other calves and their mothers in the deeper waters of the main lagoon. Here began the play between youngsters that is such an important part of the early training in the higher mammals. In the excitement of chasing and being chased by other young whales, he learned quickly to get over his early clumsiness, and turn and twist in expert flight and pursuit, his body literally flying through the warm waters. Soon he was even throwing his whole body out of the water in daring leaps and had mastered the ancient gray whale spy-hop, forcing his body straight up above the surface to half its length. The whalers used to think this was the gray whale's method of watching them, but actually it is used when they are feeding, to cascade the waters in their mouths through their baleen or whalebone strainers. But Little Star at this time was still feeding on his mother's milk.

By mid-March his mother considered it safe to take him into the open ocean and begin the long six-thousand-mile journey to Bering Sea. Unlike the bulk of the whales, who proceeded on their northward migrations still earlier and who followed the coast line rather closely, the mothers and their young, sometimes in groups and sometimes alone, moved out into deeper waters, driven by strongly held fears of men and of killer whales (more numerous near the shores of the continent) to steer clear of such dangers.

Usually Little Star swam northward side by side with his mother in a monotonous formation, both silent for long periods while the great whale strained to hear either the sounds of ships or the hunting talk of a pack of killer whales. But sometimes Star Lady paused for a half hour to sleep, or to open her great mouth and move it in slow circles in a good feeding area, taking into her maw some of the plankton she found swimming around her in the sea, her great baleen plates keeping in her mouth only what she wanted to eat, while the water rushed out to one side and she spy-hopped, lifting her head and part of her body above the waves. At other times Little Star could not contain his youthful vigor, but swam around and over her in circles, until the signal came to come close to her side again.

A few hundred miles west of the Washington coast in May, they joined four other mothers and their young in a traveling group and Little Star began to learn the signals used by gray whales to communicate with each other and to echo-locate objects in their surroundings. The "ta-ta-ta-ta-" of the echo-location sounds or clicks bounced off the bodies of large fish or anything else in the environment solid enough, and told the whales both their exact distance from such an object and how large it was. We do not know whether other baleen whales have these echo-location systems. Maybe only the gray whales have such an aid, as they are a coastal and bottom-feeding species, often coming close to rocks, beaches and shallow waters where echo-location is needed, whereas the other baleen whales are deep ocean animals and do not need it.

A series of clicks would end with a continuous sounding "trrrrrr—" or trill. No doubt these clicks and trills were reassuring noises to Little Star, telling him all was well, but once, when his mother gave the command for silence and the whole group of whales became very still, he had an uneasy feeling something was wrong, unaware that his mother was listening intently to the distant voices and echo-location clicks of a pack of killer whales. When, instead of the killers, a school of some hundred dolphins came dashing past them, porpoising over the waves and whistling shrilly, Little Star felt his mother's great relief, but he himself only envied the high speed, upward of thirty miles an hour, and the happy-go-lucky playfulness of the lithesome dolphins.

Little Star also heard the squeaks his mother and other mothers in the group used for communication as they traveled north. The squeaks he gradually learned to associate with certain movements, as when his mother, who seemed to be the natural leader of the group, urged speed to get away from an approaching ship, or let it be known he could go and play with other small whales while she slept. At such times, one mother whale usually stayed awake and on guard, ready also to call the little ones back with an urgent squeak or a more positive smash of her great tail flukes on the ocean's surface, if they wandered too far.

In early June they wandered through the Unimak Pass between two of the Aleutian Islands into Bering Sea, and here the group began to swim faster, drawn by an ancient hunger, for all their eating to southward had been only a light breakfast followed by fasting, compared to the vast dinner they were now to indulge in. Before them lay the immense shallow shelves of the Bering and Chukchi Seas, mostly less than four hundred feet deep, while the bulk of the sea bottoms ranged from one hundred to three hundred feet, an area that had once formed a land bridge between North America and Asia.

To these shallow waters now, free of the winter ice pack, the gray whales were drawn as if to a magnet. Soon Star Lady was diving down into the depths, drawing on her reserves of air in blood and muscle and arteries as well as lungs, to dive below the surface for six to eight minutes or more. On the bottom she bulldozed along taking quantities of mud and living things, mostly tiny amphipods, into her great mouth, then surfacing to spy-hop and force out the water in cascades from the sides of her mouth, while the food was retained by the sievelike baleen. While she was underwater, Little Star and other small whales remained near the surface in the care of a nurse whale, a female especially assigned to take care of the children while the mothers hunted for food.

It would be some time before Little Star learned how to dive to such depths and catch his own food and learn the final tricky act of forcing the krill or amphipods down his throat with his tongue, using it like a giant plunger. Tons of this krill was scooped up every day that summer by Star Lady and each of the other large whales, building thicker and thicker layers of blubber as stored food in their bodies to be used as reserves during the many months they would spend to the south with less to eat during the late fall, winter and early spring. Such was the yearly rhythm of life for the gray whales, perhaps unique in all the animal kingdom.

In late September, as the ice began forcing its way south out of the chilly Arctic Ocean and down into Bering Sea through Bering Strait, the gray whales began to end their feast and prepare for the fall migration, heading for the breeding grounds in Baja California. Star Lady and Little Star moved south with them, but the mother was no longer so anxious to get to the warm lagoons as she had been the previous fall when she was pregnant. She and a group of other cows with calves proceeded more leisurely southward, pausing now and then to rest and feed, if the surface plankton was plentiful. But four hundred miles off the Washington coast they heard the talk of a pack of killer whales, the faroff pinging sound and the whistles of conversation that meant these highly intelligent and dangerous animals were on a hunt. Almost instantly, by order of Star Lady, the oldest and wisest of the mothers, the gray whale pod became absolutely silent, and the mothers formed a ring around the young ones.

When the killer whale pack arrived, their immense black dorsal fins high above the water and cutting it at a speed far greater than that of the gray whales, they swirled into attack and tried to drive the mothers away from their calves. They would bite at the fins of the large whales, or at their lips, but attempt to stay away from the great tail flukes. In the melee of the fight, their flashing white-and-black bodies and their rows of large sharp teeth made them look highly dangerous, but the mother gray whales had a trump card: they could turn their bodies very rapidly and strike at the killer whales with their flukes, sometimes with deadly aim. The dodging killers escaped such strikes most of the time, but when one of the smaller black-and-white whales was struck by one of those huge tails, it was out of the fight for quite a while and sometimes for good.

Many gray whales carry for life the scars where killer whales have bitten them, but few killer whales caught by men and dissected have shown the remains of gray whales in their stomachs. It is true that occasionally a sick or wounded gray whale becomes so frightened when caught alone that it turns upside down, giving up the fight and letting itself be torn to pieces, but most of the time, like a herd of caribou or elk, the adult gray whales can keep their lesser-sized enemies at bay, as the caribou herd does the wolves.

At last one of the young gray whales, who was obviously sick and weak, dropped too far behind its mother and was immediately attacked by two killer whale bulls. When the blood began to flow through the water and the signals of the attacked whale grew weak, Star Lady, realizing any chance of saving the young one was hopeless, gave the command to abandon it, and the grays fled south again on their great migration, leaving behind another tragedy of the sea and a well-fed pack of killers.

When, in late December the gray whale pod under Star Lady reached the safety of Scammon's Lagoon, after their long journey from the Arctic seas, it seemed almost as if they gave a great collective sigh of relief and pleasure, for they had landed in their paradise. Here in the warm shallow lagoons, free of any fear of killer whales or even man, for they were now protected by law, they could lie in the warm water for hours during the days, and sleep and sleep, automatically rising to the surface every three to seven minutes to blow two or four spouts of air and vapor as much as twenty feet into the still air. At night some would swim out through the narrow lagoon entrance to hunt for morsels of food in the shallow coastal waters, bulldozing the bottom ooze for clams and sifting the waters above for sardines, then coming to the surface to spy-hop and sieve out the mud and water from their food.

When the delightful winter in the lagoon paradise was over and the increasing warmth of the sun called the gray whales north to the Arctic seas, Star Lady and Little Star swam north again with the great spring migration. But she was no longer nursing him and he was growing increasingly independent in his search for play and food. By fall the two would be separated and he would be a follower of the big males, and finally in a few years an adolescent whale with incipient ideas of mating.

Let us leap several years ahead to the time of that mating, for it has about it several unusual and interesting features, found among no other whales and, in fact, among few other creatures. When Little Star was seven years old and no longer likely to be called little, for he was about forty feet long and weighed around thirty-five tons, he formed half of a team of two males following and courting one female as she headed in fall south from the Arctic seas, all three of them eager to reach Scammon's Lagoon. There was no direct rivalry or fighting between the two males, but there was a sense of rank. Little Star was the younger and smaller of the two, and so he tacitly admitted that the other had first rights. In a sense, he became a kind of second husband, and the relationship was similar to that rarity among mankind, polyandry, when a wife has two or more husbands. The reason is that gray whale females come into heat only once in two years, so there are only one half of them ready to mate at one time, leaving twice as many males ready to mate.

It was a time of great excitement for the three, that journey south—a kind of honeymoon, though the actual mating was not likely to take place until they reached the peaceful and private backways of the Baja California lagoons. As they migrated, they swam over and under each other, both the males and the female, often gently rubbing their bodies together, or touching lightly and then parting. Merrily they swam south, the three gyrating in a rosy little world of their own, anticipation and excitement increasing with each mile they drew closer to the lagoon paradise that was waiting for them.

When they entered the warm shallow waters of Scammon's Lagoon the female became skittish and shy. She seemed to be trying to escape her two suitors and swam away at high speed with the males in pursuit. Exhausted after a while by the game of hide-and-seek, she insisted on sleeping for a day and a night, while they hovered restlessly near. Then began the chase again, from one end of the lagoon to the other, excitement mounting. At last, in a remote area where there was a feeling of privacy, she gave signs that she was ready. The older male came more and more in contact with her, rubbing his body against hers. At last they turned onto their backs, drew close together, and slowly turned to face each other, while Little Star, also in high excitement, swam below them in an unusual desire, seemingly unselfish, to hold them up in proper position.

It must be understood that the great whales, because of their size and their small eyes, find mating difficult, often failing many times before success is at last achieved, and some of them never succeed. The gray whales, perhaps alone, seem to have developed a very remarkable method of cooperation between males to make mating successful. It is at the opposite pole from the jealous guarding of his harem by the great master bull of the sperm whales.

We can be assured that Little Star in turn would have a chance to mate with his female, and that in time, as he grew older and bigger and more experienced, he would probably become a successful husband!

TOP TO BOTTOM: Little Piked Whale—also called Sharp-headed Finner Whale, Lesser Rorqual, or Minke Whale [*Balaenoptera acutoros-strata*]; Sei Whale [*Balaenoptera borealis*]; Fin Whale— also called Finbacked Whale, Finback, or Common Rorqual [*Balaenoptera physalus*]; Blue Whale [*Balaenoptera musculus*]

BALAENOPTERIDAE:

Fin Whales, or Rorquals

Very large (105 feet) to small (twenty-five feet) baleen whales, all having small dorsal fins, usually quite near the tail end, long furrows across underside of throat and breast, rather short baleen plates (less than three and a half feet long), and pointed rather than rounded side fins. All of them migrate regularly over great distances.

Fin Whale

[Balaenoptera physalus]

Also called Finbacked Whale, Finback, or Common Rorqual

Description About sixty to eighty feet long, dynamically and gracefully shaped for speed; distinctively V-shaped flat top to head; very tiny dorsal fin far down back near tail;

colored brownish gray or black above, whitish below, except lower jaw which is distinctively colored gray on left side and white on right. Both the side fins and tail flukes are dark above and white below. Baleen is usually blue gray, but forepart of right side plates are yellowish white. This is the second-largest living whale.

Range and habitat Once a common whale off the Pacific Coast, but whalers have cut down its numbers rapidly. Usually migrates south toward the Mexican west coast, off which it spends its winter, then north to spend its summer feeding in the Bering Sea area.

Habits Appears often in schools of three to ten, but more rarely up to one hundred or more, such larger schools being found at the northern feeding grounds. Occasionally it is solitary. A swift swimmer, averaging eleven to fifteen miles an hour, but under stress, up to twenty or more. The usual vertical blow (if no wind) appears like an inverted cone, reaching up to eleven to twenty feet. As it dives the dorsal fin appears, but hardly ever the flukes. Can dive for as long as half an hour; usually takes one deep dive, then several short ones, staying on surface about two minutes after a deep one to breathe. It rarely does any leaping out of water. Feeds on krill, one to two and a half inches long, in the northern seas. Like other large baleen whales, it defends itself against killer whales by its immense tail flukes, capable of giving a crushing blow.

Captain Jacques Cousteau claims that a young finback whale eats as much as three and a half tons of plankton a day, while the adults consume from one to one and a half tons. This whale tilts its head to the right while feeding, which may explain the two different colors of the jaw, the gray above water as it feeds and the white of the right side below.

Breeding in winter probably takes place around 150 to 200 miles off the Baja California coast, the newborn whales requiring warm water because of their lack of blubber. The gestation period is about eleven months, with the calves suckled for around six months. Twenty to twenty-two feet long at birth, growing to about thirty-nine feet by time of weaning; sexual maturity at sixty-three feet in males, sixty-five feet in females. May live for thirty to forty years. In the north fills mouth with krill to swallow, but also feeds on schools of small fish such as the herring.

Sei Whale

[Balaenoptera borealis]

Description   Length to sixty feet, but most seen are from forty to fifty feet. Has a beautifully symmetrical body like the fin whale, but the grooves on throat and chest do not reach as far back on body, and the dorsal fin is somewhat larger, more pointed and recurved. Color bluish gray above, gradually becoming pinkish or white below, but not on undersides of flippers, flukes or the tail end of body; baleen black with white curly hairs at end.

Range and habitat   Found in the open ocean from central Baja California north to the southern fringe of the Bering Sea, usually staying well away from the ice packs.

Habits   Generally travels in small schools, though these get bigger by combination with others at the feeding grounds in the north Pacific. Swiftest swimmer of the large whales, as fast as thirty and even thirty-five miles an hour, so can often escape killer whales. Rises to the surface in slanting fashion so nose breaks water first. The blow (or breath) extends upward six to eleven feet and frequently has with it a loud whistle. Rarely leaps up from water. Feeds by fast skimming of water for copepods and other tiny krill, covering much more ground at a time than most other rorquals. It needs to do this as the copepods are more scattered than the larger euphausids fed on by the blue and fin whales and others.

Little Piked Whale

[Balaenoptera acutorostrata]

Also called Sharp-headed Finner
Whale, Lesser Rorqual, or
Minke Whale

Description   Length up to about thirty feet; the smallest rorqual, specially distinguished by large white patch on each foreflipper and by the pure white or yellowish white baleen about ten inches long; body dark blue gray above and white below; underparts of flukes and foreflippers white.

Range and habitat   Found from southern Baja California coast (winter) to Bering Sea and Arctic Ocean (summer), usually close to coast.

Habits   Lives alone or in small schools, but may form larger schools when feeding in northern waters. Makes four to eight shallow dives before making a deep dive of three to six minutes, showing no tail when it dives; occasionally leaps out of water. Blow is small and too vague to be seen clearly. Most active at night. Feeds on krill, but also schools of small fish. Hunts deeper into ice packs than other baleen whales.

Blue Whale

[Balaenoptera musculus]

Description   Length up to 105 feet, a few weighing possibly more than 150 tons, making it the largest of all animals, past or of today. Distinguished also by the bluish, mottled with gray, upper parts, the very tiny dorsal fin far back near the tail, and the often yellow-tinged grooves on the lower jaw and front sides; the bluish gray flippers are white below. The black baleen is near to three feet long. Sometimes called "sulfur bottom whale" because of yellow bloom that appears on belly in summer in the northern seas, caused by diatoms that grow there in summer.

Range and habitat   A badly endangered species, with possibly only 1,000 or less left in the world, so that to see one along the Pacific Coast would be an event worth telling and retelling all your life! Two hundred years ago thousands moved up and down our coastal waters in migration, going north in spring to the Bering Sea to eat tons of krill, and south in late fall to the warm ocean off Mexico to breed and raise young. One or two together would probably be the most seen now.

Habits   A very shy whale, moving usually in tiny schools of two to three or alone. But this shyness is probably the result of as ruthless a slaughter of a wonderful, sensitive and majestic animal as has ever taken place in the history of the world. To the blue whale, indeed, man must seem to be the most frightening of demons, killing and killing without mercy. Average speed about six to twelve miles an hour, but can do up to twenty-four under stress. When feeding, usually makes about twelve to sixteen shallow dives to catch krill or schools of small fish, then one deep dive of about ten to twenty minutes' duration.

The blow from the blowhole is from ten to twenty-five feet high, unusually slender-looking, feather-shaped. Tail seen rarely, only just before doing a deep dive. Migrates north to Bering Sea to feed on krill in summer, and south to tropical waters in late fall when young are born. Gestation period between ten and twelve months. Calves at birth twenty-four feet long, reaching about fifty feet in three years. Calves are given milk for about seven months. Sexual maturity is reached when around seventy-five feet in length.

Whether the supposedly complete protection of the blue whale now agreed upon by leading nations will assure the future of this interesting and magnificent animal is very doubtful. If the remaining few die off without renewal, mankind will have lost a treasure of historical and social value, worth many times the millions of dollars in profits made from their slaughter.

Humpbacked Whale

[Megaptera noveangliae]

Description Length up to fifty-four feet in females, fifty-two feet in males; average weight about thirty tons. Color generally blackish above, and variably grayish to white with black splotches below. Its immense forward flippers are mainly white but trimmed above with black, their length from one quarter to one third the total length. Knobs appear on the front part of the head and lower jaw, and slightly larger lumps are found along the rear part of the back just behind the small and triangular (sometimes almost absent) dorsal fin. The black baleen is nearly three feet long. The body is thicker and not so streamlined as in most whales.

Range and habitats   Found from about twenty degrees north latitude along the Mexican coast north to the Bering Sea and sometimes into the Arctic Ocean where it migrates in summer; generally prefers coastal waters to the deep seas.

Habits   Among the slowest of the large whales, moving usually from three to five miles an hour, but capable of bursts of speed up to ten miles. It moves through the water more erratically than other whales, often appearing to be playing, its large forward fins allowing it to maneuver gracefully in several directions. It uses them also for extra speed when it "breaches" or leaps clear of the water, generally curving over to land back down in the water; it also may strike the water with its strong flukes (called "lobtailing"), creating a loud slap. The spout rises about ten to twenty feet, generally not more than twelve, appearing in a broad pear-shape, usually straight up. Food is mainly krill, taken by opening the mouth where they are plentiful, and gulping in a large mass of them. Then the head is pushed above water, the mouth partly opened and the water sieved out through the baleen, leaving the krill trapped inside where gravity and the large tongue, used as a plunger, force the food down the throat. Often small fish are sucked in also and swallowed, or the whale may open its mouth deliberately to take in part of a school of small fish when they are swimming very close together.

A Life Story

PRELUDE

The humpbacked whale has so many admirable qualities, despite its rather ugly appearance, that those who really appreciate it are infuriated to remember the great slaughter of these benign creatures by man. In several

Humpbacked Whale [*Megaptera noveangliae*]

parts of the world they have been made nearly extinct. In the first place they are probably the most magnificent singers of the sea, their deep voices moaning and booming, gurgling and fluting through the waters like a great symphony when several get together. They are also clever acrobats, leaping and whirling their bodies in play and in lovemaking, show high qualities of sensitivity and courage in helping each other in danger, and seem very curious and gentle toward men in boats who are not trying to kill or harm them. In fact Captain Cousteau of the famous sea-exploration boat *Calypso* tells stories of how they even lift their great foreflippers over a swimming man, so as not to hit him in the water. Some of his men were even pulled through the water by them! Among all the Mysticetes, the suborder of baleen whales, they appear the most likely to have something like a true language, as they communicate with each other with an unusual number of moans, squeaks, whistles, slaps, wheezes, and other sounds.

For centuries the slow speed of the humpbacks, usually less than five miles an hour, their gentleness and curiosity toward man, and their habit of migration near the coast, have made them a special victim to be attacked for their flesh and oil. Even primitive men, such as the Makah Indians of northwestern Washington in the old days, and the ancient Basques, Japanese and Filipinos, attacked and killed them from large canoes and other small boats, even in the days when only flint- or bone-tipped harpoons were used. When modern whaling ships started after them, the slaughter became more and more intense, until whalers in recent years stopped hunting them not from any real consideration but because there were too few to make them worth going after.

As in the case of the gray whales, many who hunted the humpbacks recognized the great love the mother had for her calf, and attacked the calves first with their harpoons, forcing the mother to follow them to where they could kill her most easily. The whalers also noticed that among humpbacks, unlike among most other whales, not only would the male come to help the female when she was harpooned, but also the female would come to help the male, thus making it easier to kill both heartlessly.

In studying the lives of humpback whales I was intrigued to find that often a family of them, father, mother and one or two youngsters, would be found traveling together. However, I found out later that this whale has a rather mixed pattern of life, sometimes traveling in family groups or with larger schools of up to a hundred, but at other times during migration moving in schools of males only or females only, or of females with their calves (usually more fully grown than those found in family groups). Often females without calves travel with females who have calves to act as nursemaids or aunts, to protect the little ones.

It is interesting that the two annual migrations of the humpbacked whales are undertaken almost entirely without food. One is south in the fall to escape the northern winter and to find a place in the warm seas off Mexico for breeding and calving; the other is in the spring when the whales go north to reach the feeding grounds in the cold northern waters. Food is occasionally found along the way, but is usually so scarce that the whales live off their blubber. Whalers catching humpbacks on the way north in the spring usually find they have as much as five to six barrels less per whale of oil that can be rendered from their blubber than in the same whales traveling south from the feeding grounds in fall. The humpbacks stuff themselves daily with as much as a ton and a half of krill at the feeding grounds in summer and early fall, keeping this up until they have built up enough spare blubber under their tough hides to hold them for almost the rest of the year.

It is almost impossible for us to imagine the denseness

of living krill in certain areas of the cold northern seas in summer. They are found from the sea surface down to five hundred feet or more, but are mostly concentrated in the 50-to-300-foot range, and in places where two under-sea currents, one warmer and the other colder, come together. At such places fantastic numbers of tiny algae plants also find optimum living conditions as they are touched by the sunlight, and on them feed other vast quantities of living creatures, including the krill. In such places, finding billions moving through the waters, the humpback and other baleen whales have but to open their enormous mouths and suck in the food a hundred pounds or more at a time. It sounds like a lazy whale's paradise, but we must remember that each whale has traveled over three thousand miles, almost without food on the way, to reach this banquet, and that he will have to travel the same distance back in the late fall to his breeding grounds. He thus has to make a truly tremendous effort to reach this paradise!

A FAMILY CHRONICLE

A family story must always start with a courtship. Humpbacked whales have very large brains, and show strong feelings toward one another that astonish even scientists.

Novo was a seven-year-old bull humpback with an odd habit of standing up on his fast-moving tail at the ocean's surface, with his body half out of the water, and looking around at the seabirds and whatever else was visible. Having much larger eyes than a sperm whale, he could see farther and better, but probably not as well as a dolphin, who can see out of water almost as well as under water. Peering rather myopically over the waves in two directions from the eyes on each side of his head, he would occasionally spot a ship. Then he would drop back immediately into the waves, and head in the opposite direction, a natural reaction for a large sea animal whose species has been constantly and ruthlessly attacked from such ships. Perhaps he had at one time seen one of his friends speared from a ship, or dragged into its gaping mouth by long cables, and, if his fellow whales communicated at all intelligibly, they would be full of tales of such disasters.

But on the day this story begins he had his head under the waters of the Bering Sea, about two hundred miles off the coast of Alaska, listening to some of the finest animal music in the world. It was the "sing" of about five combined herds of unattached male and female humpbacked whales, filling the sea with music before they started from the cold north on their late fall migration to the warm water mating and calving grounds off the west coast of Mexico. All were leaving with full bellies, and deeply content with the thick rolls of fine blubber under their skins, after five months or more of feasting in those Arctic waters.

If you had been there listening with him, seeing about you the green limpid waters lit by the sinking October Arctic sun, you would have felt the magic of the deep sea as never before. About you in the water would be the fantastic shapes of the some of the great whales, like monstrous dark torpedoes moving sleepily on their great white wings, their foreflippers, while the huge blunt barnacle-covered heads were lifted to listen. And from the depths all about you would come truly fantastic and weird but also beautiful sounds, long rising and falling whines and wails, liquid rumbling moans, the pulsing beats of peculiar barking yelps, long growling snoring snarls, sudden booming bellows, perhaps a long humming sputtering sound like a motorboat starting up in the distance, followed by violinlike twangings and then the sharp rasping tones of a monster Jew's harp. Now again it changed to something mysterious and lovely, almost like the angel-like voice of some woman, a voice of great richness echoing from a deep sea cavern. The sound seemed to come from very far away, with ripples and waves in the tones; and then the ethereal humming of little whispering violin notes began once more.

Now all about him in the fading light of evening, on the waters with their soft golden glow, the whales began a slow but delightful dance in which Novo joined. He lifted his great flippers like white wings and swirled his body through the waters, the hindflukes beating at first languidly and then with greater urgency, up and down, as a deep belling sounded, like stags in a forest challenging each other. But each bell had in it also the liquid water sound of the great sea, the up and down of wave beats, the echo of vast depths.

Song conversations were going back and forth through the sea as he danced, voices at different wavelengths, crossing and recrossing, but somehow not jangling; little trills, squeaks, hinges creaking, soft moans, and suddenly Novo was communicating with another whale, a female, he knew from the soft sound of her voice, infinitely inviting. She had picked his voice out from all the others and was calling. Eagerly he stroked his huge flukes in the direction of her voice, until he saw her looming out of the darkness.

Since hearing is far more important to whales than any other sense, perhaps the two recognized each other almost entirely through sound. They drew near each other and touched flippers, just the tips, ever so lightly, and it was like the giving of an engagement ring; for after that they traveled together as one, swimming down out of the cold north toward the warm and glowing tropics. Can you imagine the sensations and feelings that passed between them as they swam south? You could not tell from the expression on their faces, for whales do not express themselves that way. They express a feeling of tenderness and love by soft sounds, by touch and by their movements. How close these feelings are to ours we can only guess, but we do know they have the same or similar kinds of hearts, blood vessels, brains and nervous systems, even though they are adapted to living in such a different environment from us, the sea.

So Novo found Mega and the two moved with six other pairs of humpbacked whales, south from Alaska at first along the coasts of the Baranoff Islands, then past the snow-tipped mountains of the Queen Charlotte Isles, and finally along the long dark shape of Vancouver Island. As they traveled the two swam above and below each other and side by side always in the wonderful, wobbly, twisting, spiraling way of the humpbacked whales that seems like a dance of delight in each other under the singing waves and the song of the wind. They swam more by night than by day, often floating in the daylight on the surface, their blowholes just high enough to be above the wash of the waves, their flippers touching and then separating as they slept, their automatic nervous systems somehow keeping them from bumping. It was a sweet awakening for Novo to open his eyes (a whale has eyelids) and see Mega sleeping beside him. He would sometimes touch her lightly to waken her, and then south they would swim together again, with the cries of the seabirds ringing shrilly above them, and sometimes the distant roaring of northern sea lions sounding from rocky islets as they passed.

He was shy for a long time about touching her very closely, so it was she who finally dipped her body to his as she wafted through the water above him. By the time they were opposite Sonoma County, California, in mid-November, they were hungry enough to stop for a while and feed on the plankton swept up out of the depths by an upwelling of the Japan Curren, and they opened their huge mouths to suck in shrimps and amphipods, and occasional scatterings of small fish fifty feet or more under the surface. It was nothing like the great masses of living food they had known in Bering Sea. After the feeding he began to respond to her touches, just lightly, shyly, his body brushing against hers, now from above, now from below. But the caresses were electric and the world about them filled with dancing lights, the green water glorious with sunbeams, the fish that darted by them flashing like jewels of many colors.

As they came close to the tip of Baja California, they moved more and more slowly, almost dreamily, wrapped up in each other, until somewhere in the shallow waters near the entrance to the Gulf of California they finally mated, rising up out of the waters together, body to body, flippers wrapped about each other, the act over in a few seconds, though followed by many repetitions, sometimes at an angle, and sometimes flat along the surface.

It was not long after their last mating that some instinct told Mega she must leave Novo for a while, and she swam away to rest by herself in the warm waters. Later she searched for some small fish or shrimps or other creatures concentrated enough to warrant opening her huge mouth to take them in. It was poor pickings compared to the Arctic seas, and so like other pregnant females she became anxious to get back to the north where there would be food enough to prepare the fetus for the day of birth. The pregnant females do not think of this, of course, but the knowledge is deep within them and is why they are the first large group to head back for the north in early spring. With them traveled some groups of subadult humpbacks, teen-agers we can call them, like all of their age impatient for new adventures; while far behind mothers and fathers with newborn babies moved much more sedately, for the calves did not yet have enough blubber on their bodies to insulate them from the icy waters of the north.

Novo was meanwhile traveling far out in the Pacific with other temporarily free bulls, exploring such places as the waters around the Hawaiian Islands, and seeking that strange meeting of upwelling warm currents and cold currents where fast-growing algae attracts animal plankton in masses, on which the whales feed. Occasionally they were lucky in getting a few large mouthfuls, and these kept them going, and sustained their interest in the search, until the time came when they would hurry north for the more reliable banquets of Bering Sea.

By December Novo and Mega were brought in touch again by the deep singing of their voices through the waters just west of Alaska, but there was no anticipation of a mating now, only warm memories and a mysterious bond that would grow closer when a new life came. Meanwhile they sometimes fed together and sometimes sang together, and also sometimes wandered off with friends. When feeding together, they would pause at the surface for a few minutes of breathing deeply, storing up the oxygen supply needed for their deep diving. Then up would come their great flukes in the air, and curving over would go the huge dark bodies, and down through the gradually dimming depths to the thickest part of the krill feast where the tiny crustaceans swam through the water in their millions, feeding on still smaller creatures or on tiny plants.

The krill were not always so near the surface, rising and falling as their food animals and plants rose or fell. One time Novo dove and dove seeking for them, reaching over 1,200 feet below the surface, about as deep as humpbacks can go, before finding a large pocket of krill and getting enough for a good mouthful. Generally, by listening to the songs and voices of other humpbacks, moaning and humming through the waters, he and Mega could tell where the hunting was good and join the successful ones.

So did the six months of summer and fall pass in a frenzy of gorging, until the blubber rolled many inches thick under their tough skins. At last the increasing cold of the water read them a warning that they had better move southward again to the mating and calving grounds.

But the desire to move came at different intervals, not only to Novo and Mega, but also to others of the feeding humpbacks. The sexually immature males and females, the teen-agers, moved south first, apparently eager for warmer waters. These were followed closely by or mixed with the females with newly weaned calves, large enough to take care of themselves. Then came the mature males and the as-yet-unmated females, getting ready for new matings near the calving grounds. Novo may have traveled with these, or more likely gone on exploring trips with other unattached bulls farther out in the Pacific, temporarily lured again by Hawaii perhaps. But though we have lost him for a time, we shall soon find him when he is needed.

Mega knew she was far along in pregnancy now and she needed not a bull to be with her when her term came, but another cow. The baby inside her was beginning to thump at her body walls and give warning, and she desired the company of her kind only, other females near their times of giving birth and those odd unmated cows who appear as midwives, helpful aunts or nurses when new babies are born. Mega was well prepared for her ordeal, for she had built up fat and blubber to many inches thick under her hide, and was in the prime of health, despite the barnacles, whale lice and other free-riders and parasites that clung in the creases of her hide and the bumps on her nose.

After about two months or so of travel, she reached the shallow waters of the Gulf of California, and sought a cove, hidden by cliffs but with deep water, in which to have her baby. Perhaps her call asking for help rang through the sea, for an older female came just in time to act as midwife, an ancient whale custom. She also acted as protector against sharks and killer whales, and it was either she or Mega who sent the special pleading call, deep-toned and far-reaching, which rang through the waters for Novo. He also must come, but at the right time. Somewhere along the far sea lanes, he heard the cry, or perhaps their message was relayed by other humpbacks. So he lobtailed his flukes with a mighty crash and started swimming toward his mate. Some scientists say that such low-frequency signals may carry underwater for as many as a hundred miles.

The baby was born tail first in the usual whale way, and was not long out of its mother and helped to the surface of the water by its "aunt" when Novo came, tail-flapping and flipper-whirling his way from the deep Pacific where he had been trying probably to scare up some of the scarce food of the tropical waters.

When Novo arrived he nosed and flipper-caressed his mate and her baby, a little female, talking to them in low moans and whines that perhaps sang the praises of Mega's and his child, but certainly recalled in voice and touch the days of their courtship, and the deep feeling of one-ness that had been theirs. Humpbacks may be among the most peaceable of all whales, and there are few fights among the bulls. But Novo had not come to join his little family just to be sociable; he had come to protect them from possible enemies in the sea.

The first test came when the baby was only a week old, but had already added another two feet to her birth length of fifteen feet and almost doubled her birth weight of 1,500 pounds. The jets of rich high-protein milk that Mega shot into her mouth about every hour had added new blubber rapidly, and strengthened her muscles to be able to swim almost as fast as her parents' average four-to-five-mile speed.

Nita, as we shall call the new whale calf, was more adventurous than most, and that day she swam away from her sleeping mother and aunt to investigate a large fish. Unfortunately the fish was a fourteen-foot blue shark, almost as heavy as the baby whale and a good deal more deadly. The shark lazily turned toward the baby, its eyes becoming as cold as marble, its mouth opening to show the long sharp teeth, and its mind preparing for the sudden dash and ripping slash that would maim and kill.

Just as it started its assault, it was seen by Novo, coming in from the sea to the sheltered cove, and he wasted no time. His harsh whistle of alarm and an even louder raucous wheezing sound from his blowhole just before he slipped underwater wakened and alarmed the mother and aunt and made the baby turn suddenly. The shark's first slash with its gleaming jaws thus missed Nita's throat but tore a chunk out of her right foreflipper. In the next instant Novo's immense head butted the shark on the side next to the gills, the impact coming at better than ten miles an hour, and the shark twisted in agony and rage. It struck at his shoulder, but the skin there was so tough that the teeth glanced off it. At the same moment Novo was turning with great swishes of his immense flippers, bringing his rear end around so that he could strike with his vast tail flukes a blow that smashed onto the shark's head and drove it twisting deeper into the sea, trailing blood from its nose. Drawn by the smell, other sharks were soon ripping and tearing at the lifeless body, while little Nita with her parents and aunt swam away to safer waters, where the rising tide would wash the dripping blood from the baby's fin. The bellowing Novo then directed toward the two females after this near-tragedy probably accused them both of having been asleep when Nita went to visit the shark. Whale and human families may be a lot alike.

The baby was about six weeks old and well over twenty feet long, when about the first of March the family started its trip north, traveling with other humpbacked whale families toward the ancient feeding grounds in Bering Sea. There was enough fat and blubber insulation now under Nita's skin to prepare her for the cold water of the north. But the family was in no hurry to subject her to too much cold, and they took most of March just following the coast of Baja California, five to fifty miles out to sea. They were in the Pacific, just northwest of San Diego and in sight of San Clemente Island, when Novo gave the warning signal that killer whales were approaching. He had no powerful teeth to fight these enemies, but he did have his flippers and his flukes and of course his far greater weight.

The two females drew together on either side of the baby, while Novo followed close, keeping between them and the approach of the killers, heralded in advance by the clicking sound of their sonar. He made a few swift turns, using flukes and giant flippers to swirl his body left and right, as a boxer does when he feints left and right to catch the feel before a fight begins.

The pack of killers came crashing through the waves, the giant triangular fins of the four bulls rising four feet or more above water, as ominous as the black skull-and-crossbone flags of pirate ships. One bull was over thirty feet long, the others a few feet shorter, while the five females and three immature whales were near to twenty feet, their dorsal fins much lower and more curved.

Like a well-trained team, the killer pack closed in on the little family—but Novo was ready for them. He knew he must get to the biggest bull first, but dared not divulge the direction of his blow. So he made a strong feint to strike at a lesser bull, and as the leading killer tried to end-run him to strike at the calf, Novo made a lightning switch of his whole body by one great sweep of his fore-flippers, brought his flukes out of the water and smashed down on the thirty-foot killer's back with terrific force. The thicker, tougher skin of the humpback could stand such a blow much better than the more delicate skin of the killer, and the whole black back was laid open as by a knife cut, while the spinal column itself took a numbing, cracking shock that made the big killer whale writhe over and over in the water, his cries for help clicking and wailing through the sea. At the same instant, Mega, who had left her baby with the aunt when she saw her mate in danger, had swung up beside him and struck with a blow of her large flipper one of the other killer bulls, who was about to seize and slash Novo's lips.

This double attack was only too effective. The killer whales were already deeply shaken by their leader's cries, and pushing and shoving, they moved to get him out of danger, forgetting their own plan of attack completely. The family was saved. The killer leader would live, but he would be far more careful thereafter about attacking humpbacked whales.

The little family swam peacefully north again, watchful and alert, and worrying also about ships they occasionally saw. They were still unaware that mankind had declared a moratorium on killing them because there were so few left! People in launches and small yachts would see them on their way north as they wallowed through the blue green sea, and most of these two-legged creatures realized that even this single humpbacked whale family was one of the greatest and rarest treasures of the ocean!

FAMILY

BALAENIDAE:

Right Whales

*These whales grow up to seventy feet in length,
and are distinguished by having no dorsal fin and by
their grotesque-looking enormous heads, usually about
one-quarter their total length, with baleen plates of an
extraordinary eight feet in length. They were called
right whales by the early whalers because their clumsiness
and generally slow speed and their inoffensive natures
made them so easy to kill. Like the blue whale, they
have been almost completely exterminated.*

Right Whale

[Eubalaena glacialis]

Description Length forty-five to sixty feet, occasionally to
seventy; distinguished by the strongly arched lower jaw, with
eight to ten feet of blackish baleen plates hanging down
from upper jaw. Color velvety black over most of body,
except for a yellowish white horny material or bonnet that
covers the front of the head (usually swarming with para-
sitic crustaceans or whale lice). Some white or yellowish

TOP: Right Whale [*Eubalaena glacialis*]; BOTTOM: Bowhead Whale [*Balaena mysticetus*]

white marks or encrustations are found on the lower jaw and chin.

Range and habitat   These whales used to be in great numbers on our coast, in schools of a hundred or more, but now they are very rare migrants, in pairs or alone, traveling north to the feeding grounds in Bering Sea or the Arctic Ocean in late spring, and moving back south to the mating and calving grounds near Baja California in late fall.

Habits   Faster than its closest relative, the bowhead, it travels around six to eight miles an hour. Takes deep dive of about fifteen to twenty minutes, followed by five to six shallow dives, when feeding. When diving to feed, the flukes are lifted high above the water. Sometimes it has been seen to leap out of the water.

This whale is now supposedly protected by international law from whaling, but we will have to be watchful to make sure it is not destroyed. Like the blue whale, it is an irreplaceable treasure of immense interest to mankind.

Bowhead Whale

[Balaena mysticetus]

Also called Greenland
Right Whale

Description   Length forty-seven to sixty-two feet, rarely to seventy feet. Distinguished from very similar right whale by a bump or sharp upward and downward curve on top of head, no whitish yellow bonnet, but instead a large whitish area on front of lower jaw, and a grayish area at base of tail, otherwise completely black. Young bowheads are all black. Black baleen plates of adults are up to eleven feet long (the largest of any whale).

Range and habitat   A rare whale in the Bering Sea, seldom coming south of sixty-four degrees latitude, but living among ice floes and where there are enough open leads between ice to rise for breathing.

Habits   A very shy and timid whale, partly because of its slow speed and clumsy movements and partly because it has been hunted almost to extinction. It used to be found in small schools of thirty to fifty individuals, but is now likely to be limited to two or three, or alone. Likes to live near and in drift ice, possibly because the ice gives some protection against killer whales. Can make deep dives of fifteen to twenty minutes and has been known to dive for longer periods, possibly even as much as an hour, and down to as low as 3,500 feet below the surface, thus perhaps equalling the sperm whale. The double blow shows a sharp V against the sky, up to ten to fifteen feet high, usually directed in both a forward and sideways direction. The tail flukes are thrown up above the sea when it makes a deep dive. Feeds on tiny krill or plankton (including sea butterflies) by scooping masses of them up into its huge mouth, then forcing water out through the baleen. Herman Melville, in **Moby Dick,** tells about hearing and seeing right whales browsing on "vast meadows of brit" (the name for swarms of copepods in the Arctic seas) and producing "a strange cutting sound; and leaving behind them endless swaths of blue upon the yellow sea."

Think what a magnificent sight this would be for us and our children, but it will come back only if these whales are protected from ruthless men.

CLASS

REPTILIA

(Reptiles)

The animals of this large class are distinguished by having a dry scaly skin, a three-chambered heart whose blood stays near the same temperature as that of the surrounding atmosphere, and usually five-toed feet and claws. Reptiles include lizards, snakes and turtles. In our area only five are seagoing, and all of those five are turtles.

ORDER

CHELONIA, OR TESTUDINA

(Turtles and Tortoises)

All are characterized by a hard horny shield, or carapace, over their bodies for protection against enemies.

CHELONIIDAE:

Marine Turtles

Distinguished by having a usually heart-shaped carapace covered with horny scutes. The front legs are paddle-shaped with very long fingers. The common turtle's ability to pull the neck in under the carapace has been lost. Teeth are found only in the front half of the lower jaw, but the upper jaw has a horny plate for shearing, like a beak.

Pacific Green Turtle

[*Chelonia mydas agassizii*]

Description   A middle-sized to large (thirty- to sixty-two-inches-long) turtle, with very large swimming foreflippers. It has a lower jaw with serrated (toothlike) edges, and two prefrontal scales on the head back of the nose. There is no central keel on the carapace, which has serrated edges in the rear. The scutes or large squares of the carapace are colored brownish to olive, and often have a radiating, mottled or wavy pattern. The plastron (large shield underneath the body) is pure yellowish or white; the skin is brownish or more rarely black or gray, and some of the head scales often have yellow margins. In males both carapace and plastron are more narrowly tapered at the rear end than in females. The male also has a long curved claw on its fore-flipper, which is used to hold the female during mating. The tail of the male is much longer than that of the female, and can be curved upward or downward in a prehensile manner and is tipped by a strong spine, also used in mating to cling firmly to the back of the female.

Newborn turtles or hatchlings have a dark green to brown keeled carapace, which may be mottled and with a light border; the uniform yellow or white plastron has two longitudinal ridges. The flippers are yellowish on their borders and the rest of the skin is blackish. Hatchlings are about two inches long and usually weigh a little less than an ounce.

Range, habitats and habits   Found in the ocean and occasionally along our coast from southern California south to Chile, usually preferring comparatively warm waters. This turtle sometimes migrates across the open sea to islands, finding its way by a method of navigation we are not able to understand yet, but which is very accurate. Eggs are generally laid at night deep in beach sand far above high-water mark. It is the only marine turtle of our area commonly found basking on rocks near the edge of the sea.

Locomotion, feeding and nesting behavior   Locomotion is done mainly by the large front flippers stroking backward, the back flippers acting more as steering oars. It is a tireless long-distance swimmer, but generally travels at a speed of about one mile an hour.

Adults feed mainly on plants near beaches, particularly seaweeds and eelgrass, but also eat small crustaceans, molluscs, jellyfish and sponges. The young are more carnivorous, preying on the smaller life of the sea and shore. Because the adults are mainly plant eaters, they lack vitamins, and may try to replenish these by basking in the warm sun.

Pacific Green Turtle [*Chelonia mydas agassizii*]

Once every two to four years a mother turtle swims shoreward to some uninhabited beach at night, cautiously investigates the beach by smell, sight and taste for possible enemies, then moves higher, usually behind the first line of seashore plants, to find a good place to lay eggs. Using her flippers to dig, she moves round and round in a circle and makes a pit several feet deep in which around 70 to 200 eggs are laid. After laying eggs she covers and disguises the nesting place, even making two fake nesting holes nearby. Then she heads in a beeline for the sea, often tired by her labor, but apparently ready soon after she enters the water to be mated by a male or males who are waiting there for her.

The nestlings dig out of the sand-covered nest anywhere from thirty to seventy-two days (more commonly forty-five to sixty) after the eggs were laid, usually coming out at night to avoid both the heat of the sun, which could kill them, and the danger of too many predators. Even so there are so many raccoons, foxes, coyotes, dogs, men and others eager to eat the eggs that many are dug out before they have a chance to hatch. After hatching, rats, cats, snakes, lizard, gulls, herons and several other creatures join the first group mentioned to hunt down and eat the hatchlings, so many being taken that few if any reach the sea. In the sea, those who survive the first predators are attacked by many carnivorous fishes. For the tiny percentage that live on to be adults, the main predators at sea are sharks and men. Nesting females may be attacked by dogs, jaguars and men.

Despite all the hungry mouths that seek them, green sea turtles are the primary source of the turtle meat and soup that reach our tables. The sad thing is that the human catchers of turtles, more anxious to get immediate profits than interested in the long-time prosperity of turtle hunting, have been catching these turtles at such a rate that they may soon be wiped out as a species. Legislation is desperately needed to protect the turtles!

Pacific Hawksbill Turtle

[Eretmochelys imbricata bissa]

Description   Small to medium in size, eighteen to thirty-seven inches, with two pairs of prefrontal scales on the head, and with dark foreflippers like the dark brownish green carapace, which is shield-shaped with straight sides in adults and heart-shaped in young turtles. The young have a tortoiseshell pattern on the carapace due partly to the forward scutes overlapping the hind ones. This overlapping gradually changes, until the scutes begin to fit neatly together as the young become adult. The yellow plastron is hingeless and flat, but in the young may have two longitudinal ridges and some dark blotches. The head is black to dark brown, but the jaws, chin and throat are yellow, the

TOP: Pacific Ridley [*Lepidochelys olivacea*]; BOTTOM: Pacific Hawksbill Turtle [*Eretmochelys imbricata bissa*]

jaws often having brown bars or streaks. The lower jaw has a smooth cutting edge, occasionally slightly serrated; on the upper jaw there is a hard and elevated vertical ridge on each side of the inner surface. The snout looks like a hawk's beak, but is not notched at the tip. Males have long thick tails, compared to the short ones of the females, long strong claws on the foreflippers, and usually a concave surface to the plastron (undershield), all used in mating.

Range, habitat and habits   Found from the California coast south to Peru, liking rocky places in shallow coastal waters, but also estuaries, bays and lagoons with mud bottoms and little vegetation; sometimes found in creeks near sea level; more rarely found in deep oceanic waters. Females come ashore to dig nests high on sandy beaches along the Mexican coast, but very rarely in California. This turtle is a vicious biter and difficult to handle, holding on like a bulldog to what it has bitten.

Locomotion, feeding and nesting behavior   Has a cruising speed of about one mile an hour, but can double this when frightened. The large foreflippers are the main aid in locomotion, acting like oars. It is an omnivorous feeder, eating various seaweeds, some fish, squid and octopi, but liking best to catch and eat various crustaceans and other medium-sized sea invertebrates. Hatchlings are mainly plant-eaters at first, becoming more omnivorous later.

Females come ashore on sand beaches to dig nest holes and lay eggs about once every three years, and are more wary than the green sea turtles, mainly coming ashore late at night and craning their necks in all directions to sniff out or see possible enemies. If disturbed they will head quickly back to the sea. Finding a good place among beach plants in sand high above the lower beach, the mother digs a deep hole and deposits her eggs (averaging 160), which incubate in about fifty-two to seventy-four days. The young hatchlings dig their way out and head for the sea, mainly at night or in the early morning. Raccoons, dogs, jaguars, man and other large animals hunt for the eggs and eat them. Rats, dogs, sand crabs and large birds eat the hatchlings on their way to the sea, and in the water predaceous fish attack them. Adults are attacked sometimes by sharks, but man, who especially wants the translucent scutes from the carapace to be used in the turtleshell of commerce, is the most dangerous predator. In fact, like the green sea turtle, the hawksbill is so persistently hunted that it too is in great danger of extinction. One thing we human beings can do to stop the slaughter is to refuse to buy anything made of turtleshell until the turtles are better protected.

Pacific Loggerhead Turtle

[Caretta caretta gigas]

Description   Adults twenty-eight to seventy inches in size, weighing generally from 200 to 400 pounds, but some of 1,000 pounds or more have been recorded. This is the only sea turtle with a reddish brown carapace, sometimes touched with olive and with the scutes often bordered with yellow; the undershield or plastron is cream to yellow-colored and has two lengthwise ridges that vanish with age. The head is particularly large and broad, and colored yellowish brown red to reddish or olive brown, with yellow fringed scales; the powerful jaws are yellow brown. Legs and tail are dark in the center with yellowish edges and undersides. Shells of males are generally narrower than those of females, and tails are much longer.

Range and habitats   Ranges mainly in coastal waters of the central Pacific and Indian Oceans and bordering seas, but on the Pacific Coast rarely found north of Southern California. May be found as far as 500 miles or more out to sea, but also enters lagoons, bays, salt marshes and even into the fresh waters of creeks and rivers. The female lays her eggs in a hollow she digs high on sandy beaches, then covers over with sand.

Habits   The powerful foreflippers drive these turtles through the sea at somewhat faster than a mile an hour for many hours at a time, but they are capable of turning very quickly and sharply to counter an attack, and use both flippers and powerful sharp beak very effectively as weapons. They usually breathe about every two minutes when swimming, but can stay under water at least twelve minutes and some up to twenty-five, often resting at such times on the bottom. Loggerheads are omnivorous eaters, feeding on many shellfish, squid, jellyfish, sponges, crabs, and various fish; they also eat eelgrass and seaweed, often feeding in rocky areas, about coral reefs, and near old wharfs and wrecks. Men, sharks and killer whales are the chief predators of adults at sea, but egg-laying females on the beaches are attacked by dogs, jaguars and man, while the young are eaten by a myriad of small predators, from raccoons to seabirds to snakes.

A Life Story

PRELUDE

The sea turtles are among the most ancient of all sea folk, far more ancient than even the whales and infinitely more so than the seals. In fact they are probably the first of air-breathing land vertebrates to find their way back to the sea. This is due to turtle- or tortoiselike creatures appearing so early in the history of the earth. Way back

Pacific Loggerhead Turtle [*Caretta caretta gigas*]

in the Permian period, around 260 million years ago, there appeared a creature with a bowed-up shell on its back, obviously a kind of armor for protection against predators, and so an ancestral turtle. And by the Jurassic period, about 170 million years ago, some similar creatures were beginning to turn back to Mother Sea and learn to swim in its waters. The first definite ancestors of the modern sea turtles did not appear until the Cretaceous period, about 130 million years ago, still very ancient compared to the most ancient whales of the Eocene epoch, about fifty million years ago.

Why did they go back to the sea? The time is so far back our speculation on this is little more than guesswork. Possibly it was because of drouth on land, but more likely due to some of them coming to prefer the food they found at the edge of the sea and so moving into the water to find still more of it. Though the sea turtles, down through the millions of years, became more and more streamlined to allow greater speed and maneuverability at sea, the solid hard shell they had to wear on their back always hindered speed, though giving good protection against enemies. The leatherback turtle is an exception to this rule, as it developed a leatherlike covering for its back to take the place of the bony carapace, and this made it a much more streamlined turtle with a faster speed in the sea.

Though the sea turtles went back to the sea long before the whales, they never went as far as the whales in becoming completely adapted to marine life, and this may simply be because of the conservative nature of reptiles as compared to mammals. Thus the whales learned to give birth in the sea whereas the sea turtles, like those more recent sea mammals, the seals, bear their young on land. This forced them to keep their legs, in order to be able to travel on land, something the whales have dispensed with.

Though the armor the sea turtle carries, the horny

shell and the thick leathery tough skin, does hinder its speed in the water, it also protects adult turtles from attacks by all but the very largest sharks. Even the great white shark, king of the man-killers, might find a half-ton leatherback or loggerhead a tougher customer than he bargained for, as both these turtles have beaks capable of shearing a two-by-four in two with about one slash, and their huge foreflippers can be used like swords or clubs. More than one turtle-hunting man has found them too hard to attack even with razor-sharp harpoons and they have been known to break strong rope as if it were twine!

There are three other great differences from sea turtles and sea mammals. One is that turtles are cold-blooded animals so, unlike warm-blooded mammals, cannot stand very cold sea water, which limits their range to not much farther north than southern California. The second is that they lay eggs on sandy beaches in large quantities, up to about 200 at a time, but do nothing to protect them after they are hatched. This means that large numbers are killed before they have a chance to grow up, probably little more than one percent becoming adults. The third is that sea turtles, if they can survive the very dangerous time when they are small, have a very long life span. We cannot know for sure, but we assume that a very large sea turtle of half a ton or more is at least a hundred years old and possibly a lot more. And such creatures, to live that long, must be very wise indeed in the ways of the sea!

CAGUAMA

The large female loggerhead, swimming strongly and purposefully out of the Pacific Ocean on a dark hot night in June, let the waves on a Baja California beach wash her ashore till her flippers touched the sand and she could start to drag herself ponderously up the beach. Moving in the deep darkness under a star-spangled sky on that wild and lonely beach, she stopped occasionally to peer around and sniff, to determine if any predators were near. Observing or sensing none, she continued up the beach until she had passed the highest tide mark and was coming to the first sand dunes. In the flank of one of these she began, using all her flippers, to dig a depression that would hide her body from view in any direction. When this first hollow was made so she was out of sight, she stopped using her front flippers and began scooping out sand with each hindflipper, following each scoop with a deft movement to spread the sand away from the new hole she was creating in the middle. This hole she continued to dig until it was about ten inches deep and ten inches wide. When deep enough, she laid all her flippers out flat, palms down, and inserted her rear end into the hole, in order to lay her eggs. During all the time of the digging and now the laying, tears filled and flowed from her eyes, but they were tears of neither joy nor pain, but simply a method of washing the sand thrown there by the digging out of her eyes.

Every time a group of eggs were deposited in the hole, she would lift her head and arch it, giving a soft snort or a long sigh, as if there were a little pain but also great relief at each laying. When 155 eggs had been laid, she seemed to know that that was all and began rapidly covering both holes with sand. First she used the hindflippers to draw up sand from behind her, but gradually she began to use the front flippers too, although mainly to sweep back sand where the hindflippers had dug it up, thus eventually thoroughly covering both depressions she had made with heaps of sand, pressing down the whole mass with firm pats of her hindflippers, body and head raised to exert greater pressure.

After several passes and flinging about of sand, she decided she had disguised the nest enough, raised her head high to wipe away both tears and sand with the outer hyoid membranes, and started for the sea. The trip to the sea was made with a faster than usual sliding and dragging movement that carried her straight there. She paused for a moment as the first waves touched her, and then slid down and into them without looking back.

Out beyond the breakers a group of males were waiting for her, since just after egg-laying a female is ready to be mated. Three she scornfully rejected by the female trick of swimming with body upright in the water and a facial motion that means No!; but a fourth and larger male she accepted after some initial flipper-waving and pushing. He climbed onto her back, using the long sharp curved claws that extend from the ends of all four flippers, to hang on tightly, and they were soon mated. From one or more such matings a year she would lay three or four or even more clutches of eggs on the sandy tropical beaches of the Mexican coast. This might amount to as many as a thousand eggs laid, of which possibly ten would eventually reach adulthood. Consider how different this is from the sperm whale mother who has a single child about every two years and, during those two years, teaches it to face the dangers of life!

Down under about sixteen inches of sand the 155 eggs stayed warm and snug, though not too warm, as they were deep enough to miss the scorching heat of the tropical day. About twenty-five of them had never been successfully fertilized and were already dead, but in the others life was steadily growing. In about sixty days the first little loggerhead turtles broke their shells one warm night and started burrowing through the sand. Most burrowed straight up, but a few burrowed sideways, and some of these would never reach the surface. All the little turtles were somewhat under two inches in length, with brown carapaces, marked with three longitudinal ridges. Their heads were large in comparison with their bodies, their eyes large and dark. One small turtle broke out of its unusually large shell about fifty minutes later than the first ones, and showed unusual energy and activity. He started digging sideways, but soon changed and began to work his way to the surface, moving so fast that he passed the rest and was the first to come out of the sand into the night air.

Instincts probably almost as ancient as his sea turtle family took over and he turned his head in a circle, rejecting the broken profile of the sand hills behind the beach, but accepting the wide, smooth, and more light-filled profile of the sea and its beach, and so moving down the slope toward it, his instinct telling him that a downward motion was right. It was fortunate for him that he was the first to come out of the sand, for he had no sooner disappeared from the nest site, and the second and third young turtles had broken the surface, than the black mask of a raccoon rose above the nearest sand dune, his dark eyes sparkling at the sight of little turtles leaving their nest. Soon the loggerhead turtle's worst land enemy, worse even than man, was on his way to grab himself a tasty meal of young turtles.

Caguama is the name the Mayo Indians of Sinaloa Province, in western Mexico, give the loggerhead, a name borrowed from the Black Caribs of the Caribbean coast; so we shall call our turtle Caguama. He needed a good name as well as lots of luck, as probably not more than one to three turtle babies survived to adulthood out of that nest of 155, whose mother had long ago deserted them for the sea.

Caguama was hardly six feet from the edge of the ocean when a land crab, about six inches in diameter, started after him. The huge pincer would have crushed his weak little shell in an instant had not an unusually large wave sent its final crest sweeping onto the land and over the two of them. In the tumbling water, the land crab scrambled back to dryness while the little two-inch-long loggerhead kept stubbornly ahead into the sea.

But escaping one enemy had only brought him into an environment where there were a myriad more. Lurking near him in the sea were at least a dozen fishes big enough and hungry enough to swallow him. Only his camouflaging brown and gray colors kept him hidden in the dark night of the sea, as a current swept him northward to a more rocky beach. Here he was fortunate to be swept on the lee side of an enormous rock where a deep tide pool was sheltered from the full force of the sea. There his brown shell merged with some brown seaweed that clung to the low and middle tide zones of the rock. On the seaweed he was able to eat, and also hunt around its holdfasts or rootlike attachments to the rock for tiny amphipods that he could snap up in his scarcely larger bill.

Around this and neighboring rocks and jungles of seaweeds he was able to live in comparative safety for several months, gradually growing larger until a small storm sent his now eight-inch length careening out to sea in the undertow of a big wave. Fortunately again it was nighttime when his second sea voyage began, and he floated and swirled away into the deeps, so dark himself he was all but invisible in the night waters. His swimming was only modest as yet because of his small size, but his floating ability was tremendous due to the large air bubbles under his carapace, and he had no trouble finding times to breathe even in the storm-tossed waters.

Morning found him ten miles at sea and near a large raft made by some coastal Mayo Indians and long ago abandoned in a storm. This raft, already worn and tossed by the sea, yet strongly bound by fiberlike ropes of strong sisal yucca, had long since picked up a living cargo of seaweeds and the various small living things that dwelt among them. Such a floating jungle was an ideal hunting ground for an eight-inch turtle youngster among the clinging seaweeds and the hydroid gardens of the old logs. No animals on this raft were larger than he, and the few crabs that were his equal, he could bluff by waving his large front flippers at them. All around him was the sea full of hundreds of large fish that could have swallowed him at a gulp, but on the raft he was king.

With this raft he traveled over the eastern Pacific for three years, twice so storm-tossed that he was nearly torn loose from his safe island. But the end of this ideal existence came when a steamer ploughed with horrendous grinding noises into the raft one calm night, and he was tossed into the swirl along the ship's side and swept within inches of the knifing propellers. For three years he had been protected from myriad enemies and had grown enormously to a respectable twenty-one inches in length. And now he was protected above and below by a tough bony carapace and an equally hard plastron, the leathery skin on his neck, flippers and tail being also tough enough to fend off bites from most smaller sea creatures.

However, instinct and a growing knowledge of the sea and its inhabitants, from watching them from the raft, told him that there were still plenty of enemies in the open sea and that he had better head for somewhere safer. How he knew the right way to reach the coast of Baja California, from one hundred miles out in the open sea, we may never know. It might have been the very simple knowledge that land would lie in the direction of the rising sun, because most seabirds came flying from there in the morning. At any rate, it was in this direction that he swam, the two powerful front flippers stretching over two feet across his front, and working in unison to row him through the water.

Feeling for the first time in his life the joy of traveling vigorously by his own power through wide-open space, and the ceaseless rhythm of the sea as the waves rose and fell beneath him, he swam tirelessly for some time, pausing only when he came to a small island of kelp, about eighteen-by-ten feet in size, that had broken loose from a distant rocky shore. It was in the center of this island, surrounded by the stems and leaves of the seaweed, which disguised his brown body completely, that he lay and rested for the next twenty-four hours. It was fortunate for him that he did, for two fourteen-foot-long hammerhead sharks, their jaws full of great white teeth, rose out of the dark depths of the ocean and passed close by his island on their way to some unknown destination, either one of them being easily capable of killing and eating him.

Perhaps he sensed how close he had been to death, for when he left the kelp islet, it was in the dark of night, under the slow wheeling of the bright clear stars, and his body seemed to merge with the dark waves and become one with them. In the middle of this night a school of squid rose from the depths with glowing luminescence and surfaced all around him, their bodies and tentacles trailing filmy streaks of fire through the sea. Through this mass of fire and life there soon came crashing a group of young bachelor sperm whales or cachelots, their great dark bodies sometimes blotting out half the sky for the turtle. Somehow he rode through their turbulence without being swallowed, as the whales wolfed down the squid, and some of the smaller squid he was even able to grab for himself as they fled from the whales. It was a delicious taste that he was ever after to seek for himself.

The following day he slept quietly in one place in the middle of the sea, his shell filled with just enough air to keep him buoyantly rocking with the movement of the swells, his head down for two or three minutes, then rhythmically up for its regular sucking-in of air. Such a motionless body draws far less attention from predators than one moving through the sea, and so he slept in comparative safety during the daylit hours. In the night he was again under way toward the coast, and made landfall about four hours after midnight, sculling in the dawn, his front flippers driving him through an opening that led into a lagoon on the Baja California coast. There, in warm calm waters, too shallow for big sharks, he found what he was perhaps instinctively looking for, a place to find food with little danger from large predators.

He fed here through the next eleven years on sponges, jellyfish, clams, oysters, conches, mussels, small squid, shrimp, crabs, barnacles, tunicates, sea urchins and even small fish, such as rocky fish, caught by a quick driving flash of his head and neck, shot out from his body. He would also graze among the forests of seaweeds, swimming among them and nipping off the more tender leaves, or swim up into the salt marshes and estuaries to nibble at the long, bright green, waving fronds of the eelgrass. It was an ideal life for a loggerhead of medium size, a place rarely frequented by man because it was surrounded by deserts. He even in his fourteenth year found an old shipwreck, one that lay just outside a lagoon along the ocean shore, but was protected from big waves by a headland, where he spent months scrounging through the dim rooms and deep holds where men had formerly lived and where a few pieces of bones and skulls still lay. Here he was big enough to meet in fierce combat, beak to tooth, the five-foot moray eels that hunt octopus in the old ship crannies, and tear them into edible chunks, their teeth clanging on his hard shell in their dying struggles. The octopi could slip through holes too small for him to follow, but several times he caught them when they were escaping from chasing eels backward through the waters, leaving behind them the black pseudomorphs with which they had tried to mislead their pursuers. Once he even met an octopus twice as large as himself and fought it for twenty minutes under the sea with held breath, while its sharp-as-a-razor beak tried futilely to pierce his thick undershell. At last his own beak sheared through to its brain as he forced it against a rock wall, and the battle ended.

In such hiding places many of the great sea turtles find safety to live through their middle years, unobserved by man, until at last they feel big and strong enough to venture forth on journeys of exploration into the unknown sea. Whether such urges are like those of men who go exploring, or are simply the need of a larger turtle to find larger food to fill his belly, we may never know. But Caguama in his twenty-third year weighed over three hundred pounds and had reached a length of over four-and-a-half feet, and evidently felt capable of a long voyage. In that year he left the coast for the open ocean. Such a creature is tough almost beyond belief, with a carapace and plastron of hard bone too strong even for the teeth of large sharks, a head and neck capable of striking like a rattlesnake in many directions, and a razor-edged beak that can shear through a man's arm, bone and all!

So Caguama traveled north and west that summer, staying with the warm currents, for no sea turtles except perhaps the great leatherbacks like to stay in a temperature below 40 degrees Fahrenheit. Perhaps his sea wisdom told him it was better to stay north of the usual limit of the great white shark, one of the few creatures of the sea besides the killer whale capable of cutting off his flippers with a savage bite. Five hundred miles west of San Francisco he encountered a great school of mackerel streaming through the sea in thousands of silver streaks, far faster than a sea turtle could swim. But he could twist his body like lightning and strike with neck and head and open jaws, in a blinding flash that culled a mackerel from the water above him as it drove by. He caught and swallowed four in this way before the school left him floating on the surface now, full and satisfied.

Sometimes he went without food for several days, living on his fat and moving slowly to conserve energy; but he was learning the connection between water temperatures and food. Where warm water currents meet cold, welling up from the depths or even colliding from different directions, the plankton—the almost microscopic animal and plant life of the sea—find the best conditions for growth. And where the plankton came, there followed the larger life that feeds upon it, and in turn there came the sharks, the turtles, the seals and the whales that feed upon this larger life.

Once when he was feeding amid a school of small squid, their translucent bodies shooting backwards around him, driven by jets of water from their forward siphons, a group of eight ten-foot-long tiger sharks rose out of the depths below to investigate him. Following an ancient instinct of his kind, he rose slowly to the surface and lay there floating, absolutely still, till the sharks, apparently regarding him as only some odd piece of driftwood of no food value, swam away.

It was in his twenty-ninth year, when about five feet long and weighing close to four hundred pounds, that sharks gave him the most trouble. Near the Hawaiian Islands where he had been investigating the abundant life of the coral reefs, and growing fat on eels dragged by his powerful beak out of their holes and the large shellfish he tore from the coral, a group of about fifteen-foot-long sharks came nudging about him like dark sinister torpedoes, their white razor-sharp teeth grinning in their cruel mouths. Again he floated to the surface, and held as still as possible not to arouse them, but one suddenly took a sharp bite at one of his long flippers, and he was only just fast enough to draw back and strike a blow that shook the shark to its bones. The huge blue shark whirled and again came at him, with hard and cold eyes, and its body drove with lethal intent turning sideways and holding its mouth open.

Caguama either knew instinctively that if the shark drew blood, the others would be upon him in raging bloodlust, or he had learned this wisdom from hard experience. He acted decisively, whirling his body with a deep thrust of one flipper so that the edge of his iron-hard carapace cut like a saw, square on the closing mouth of the shark. It broke the teeth off in a long string and incapacitated the crushing force of the shark's jaws. In the next moment Caguama drove his four powerful flippers down into the water with such force that his whole body was thrown up clear of the sea and down again with such a resounding splash that, both above and below the sea, it sounded like the boom of a great gun. The terrific noise, perhaps quadrupled under the water, made the five great sharks jerk away in terror and surprise. Again Caguama lifted his huge body with the power of those muscular flippers and again he smashed the surface of the sea, and away the sharks fled into the depths.

It was these tricks and an increase in his sea knowledge that kept him alive during many ensuing years, traveling south to the Galápagos Islands and even to the coast of Chile, west to the Marquesas Islands in the South Pacific and northwest to southern Japan, covering the vastness of the father of oceans as if it were his private lake. When he weighed nine hundred pounds and was nearly six-and-a-half-feet long, a monster of his kind but still filled with incredible vigor, he was attacked by five fishermen in a boat along the coast of Mexico. They tried to drive a harpoon into his neck, but he twisted about and tore it from the man's hand as easily as a child breaks a fragile toy. Two harpoons bounced off his carapace and still another found his side too tough to penetrate. Then an infuriated Caguama rose beside their rowboat, bit off an oar that was pushed at him as if were a stick of candy, and threw his whole weight at the boat, in such a way that it capsized, throwing the men into the sea. Here he could easily have torn them to pieces one by one, but, content with overthrowing them, he headed seaward, leaving behind some wiser fishermen!

And we will be wiser too if we respect the great sea turtles with their wondrous journeys over the deep blue sea, and protect them and their nests and eggs from the rapacity of those who may otherwise exterminate them. May they live forever, these tough old giants of the singing waters!

Pacific Ridley

[Lepidochelys olivacea]

Description   This rather small sea turtle ranges from twenty to twenty-eight inches long as an adult; it has two pairs of prefrontal scales on its head behind the nose, and the olive-colored carapace is heart-shaped, having its highest point just behind the neck, and its rear edge serrated. The hingeless plastron is greenish yellow to greenish white; the skin is olive-colored above and more yellowish to whitish below. The snout is short and broad, while the wide head has concave sides. The carapace of the young turtle is more oval. Males have long strong tails, used to hold the female in mating, concave plastrons to fit the back of the female, and a long powerful curved claw on each front flipper, also used in mating.

Range and habitats   Found from California south along our coast, but rarely to the north. Likes shallow and protected marine waters near land, also larger bays and lagoons. Rather rarely seen in the open ocean. Nesting is on sandy beaches but generally to the south of California.

Locomotion, food, nesting and other behavior   Swims at a rate of about one mile an hour. A very carnivorous turtle, feeding on crabs, fish, snails, sea urchins, oysters and jelly-fish. Seaweeds are more rarely eaten. Nesting and egg-laying occur anytime from September through January.

   Females dig a flask-shaped nest hole about twenty-two inches deep in the sand, high on a beach and generally at night. The nest is dug and the eggs laid in about an hour's time; then the mother heads for the sea where some males usually are waiting for her to mate again. Incubation of the eggs takes forty-nine to sixty-two days. Numerous birds and mammals rob the nest, or try to catch and eat the young as they come out of the nest and head for the sea. Sharks and men attack the adults in the sea, but this turtle has been fairly successful in using elusive behavior and camouflage to escape them.

DERMOCHELIDAE:
Leatherback Turtles

Only one species, as described below.

Pacific Leatherback Turtle

[Dermochelys coriaceae shlegelii]

Description The largest of all sea turtles and the only one with a thick leathery back and a smooth skin covering a carapace composed of many small bones to form seven narrow ridges down the back. Color generally black or dark brown, but sometimes with yellowish or whitish spots. The tough plastron (underbody shield) is whitish with five longitudinal ridges. The enormous front flippers are nearly three times as large as the hindflippers, but the whole appearance is of a more streamlined turtle and one made for higher speed travel than any other. The average adult is about five to eight feet long, weights varying from 600 to 1,000 pounds, but some giants have been reported ten feet long and weighing as much as 1,500 pounds to over a ton. The width across the front flippers varies from six to twelve feet and is always greater than the length. Juveniles appear covered with scales on the back, which disappear as adults. No claws are found on the flippers. Females have shorter tails; males much longer. Juveniles are yellow-spotted.

Range and habitat   Found throughout the Pacific and Indian Oceans, as far north as Japan and British Columbia, and as far south as southern Chile, evidently being able to withstand the effects of cold water better than any other sea turtle. However, they are quite rarely seen by man, as they prefer traveling in the deep sea. Females lay their eggs on sandy beaches. Adults rather rarely come into shallow coastal waters, and juveniles are hardly seen at all, so that where they live is still a secret of the sea.

Habits   The leatherback is an omnivorous feeder, favoring jellyfish, sea urchins, crustaceans, squids, tunicates, small to medium-sized fish, and seaweeds. Speed is probably faster than that of any other turtle, possibly as much as four miles an hour. The huge foreflippers act as oars, driving the turtles ahead in unison. Sometimes they travel in groups, especially to good feeding grounds. When the female comes ashore to nest, she pays less attention to possible danger than other sea turtles, pulling herself clumsily up the beach to a good stretch of sand above high-tide mark. Here she begins to dig first with her foreflippers, then with both fore- and hindflippers, digging a concealing pit, then the nest hole where the eggs are laid. Ninety to one hundred thirty eggs are laid and the young turtles come out of the nest around fifty-five to sixty-five days later, when they are immediately subject to attacks from many kinds of predators, from racoons and dogs on land to sharks and other large fish at sea. Few if any reach adulthood.

Leatherbacks are noted for the strange, loud and even terrifying noises they make when attacked or hurt, bellowing, roaring and groaning. Large leatherbacks can be extremely dangerous when aroused, capable of sinking small boats and biting oars in two as easily as matchsticks!

Pacific Leatherback Turtle [*Dermochelys coriaceae shlegelii*]

Though the young have many enemies, the adults have only one really serious enemy, man. Luckily they are considered poor eating, and have no carapace with valuable turtleshell, so they may have a better chance of survival than most sea turtles. Their oil is used in cosmetics and this yet may be their undoing. We should, however, protect thoroughly this most enormous and remarkable of all sea turtles, that our children and their children shall know the mystery and the wonder of them down the centuries to come.

EPILOGUE :

LAST CALL FROM THE SEA

Next time you go down to the beach or take a ride on a boat or ship out over the waves, look where the waters foam and splash over the rocks or against a ship's side, and hear them murmuring and booming. Listen to the wind sighing and singing through the trees or through the lines that hold the masts. Watch for the seabirds soaring against the blue and the sea mammals and turtles drawing their lines of ripple and foam through the waters. Try to feel the mystery and wonder of the sea, the dramatic events in myriads of lives, some of which are filled perhaps with as much sensitivity and yearning, love and joy, as the lives of humans. The stories in this book may perhaps have opened your eyes and hearts to some of these lives; but, remember, this is only a beginning, and you can go on to deeper and richer experiences with Mother Sea and her denizens, whose delight in living, but also whose agonies, can catch our hearts and make us realize —as we should realize—the oneness of all living things.

Yet a dark curtain may fall over this scene of beauty and delight, of mystery and adventure, of conflict and harmony, because mankind may be playing the last act of a drama which could end as a most terrible tragedy: the slow destruction of our living world, its strangling by the human greed and blindness that causes the pollution of even the mightiest oceans and the nightmare destruction of many of the most fascinating creatures on earth.

Will we allow this dark curtain to fall on a once green and lovely world? Or will we dedicate our lives and energies to reversing this trend and making a better tomorrow?

BIBLIOGRAPHY

GENERAL BOOKS

Anderson, Harold T. *The Biology of Marine Mammals.* New York: Academic Press, 1969.

Del Rey, Lester. *The Mysterious Sea.* Radnor, Penna.: Chilton, 1961.

Gaskin, D. E. *Whales, Dolphins and Seals.* Auckland, New Zealand: Heinemann Educational Books, 1972.

Ingles, Lloyd G. *Mammals of the Pacific States.* Stanford, Calif.: Stanford University Press, 1965.

MacGinitie, G. E., and MacGinitie, Nettie. *Natural History of Marine Animals.* New York: McGraw-Hill, 1968.

Orr, Robert T. *Marine Mammals of California.* Berkeley: University of California Press, 1972.

Wood, Forrest G. *Marine Mammals and Man: The Navy's Porpoises and Sea Lions.* Washington, D.C.: Robert B. Luce, Inc., 1973.

SEA OTTERS AND POLAR BEARS

Bruemmer, Fred. "A Day in the Life of a Polar Bear." *Natural History,* vol. 77, no. 8 (October 1968).

McCracken, Harold. *Hunters of the Stormy Sea.* Garden City, N.Y.: Doubleday, 1957.

Perry, Richard. *The World of the Polar Bear.* Seattle: University of Washington Press, 1966.

Sandegren, Finn E.; Chu, Ellen W.; and Vandevere, Judson E. "Maternal Behavior in the California Sea Otter." *Journal of Mammalogy,* vol. 54, no. 5.

Seed, Alice et al. *Sea Otter in Eastern North Pacific Waters.* Seattle: Pacific Search Books, 1972.

Stirling, Ian. "Midsummer Observations of the Behavior of Wild Polar Bears (*Ursus maritimus*)." *Canadian Journal of Zoology,* vol. 52, pp. 1191–1198 (1974).

SEALS AND SEA LIONS

Bartholomew, George A., and Colias, Nicholas E. "The Role of Vocalization in the Social Behavior of the Northern Elephant Seal." *Animal Behavior,* vol. 10, nos. 1 and 2, p. 7.

Brigg, Kenneth T. "Dentition of the Northern Elephant Seal." *Journal of Mammalogy,* vol. 55, no. 1.

Burns, John L. *The Pacific Bearded Seal.* Juneau: Alaska Department of Fish and Game, 1967.

———. "Remarks on the Distribution and Natural History of Pagophilic Pinnipeds in the Bering and Chukchi Seas." *Journal of Mammalogy,* vol. 51, no. 3 (August 1970).

Harrison, R. J.; Hubbard, Richard C.; Peterson, Richard S.; Rice, Charles E.; and Schusterman, Ronald S., eds. *The Behavior and Physiology of Pinnipeds.* New York: Appleton-Century-Crofts, 1968.

LeBoeuf, Burney J.; Ainley, David G.; and Lewis, T. James. "Elephant Seals on the Farallons. Population Structure of an Incipient Breeding Colony." *Journal of Mammalogy,* vol. 55, no. 2, p. 370.

Matthews, L. Harrison. *Sea Elephant: The Life and Death of the Elephant Seal.* London, England: Macgibbon and Kee, 1952.

Maxwell, Gavin, with Stidworthy, J., and Williams, David. *Seals of the World.* London, England: Constable, 1967.

Newby, Terrell C. "Observations on the Breeding Behavior of the Harbor Seal in the State of Washington." *Journal of Mammalogy,* vol. 54, no. 2, pp. 540–543.

Odell, Daniel K. "Seasonal Occurrence of the Northern Elephant Seal, *Mirounga angustirostris,* on San Nicholas Island, California." *Journal of Mammalogy,* no. 1, p. 81.

Perry, Richard. *The World of the Walrus.* New York: Taplinger, 1967.

Peterson, Richard S., and Bartholomew, George A. "California Sea Lion Vocalization." *Animal Behaviour,* vol. 17, no. 1.

———. "The California Sea Lion." Special Publication No. 1, The American Society of Mammalogists (December 5, 1967).

Peterson, Richard; Hubbs, Carl; Gentry, Robert L.; and DeLong, Robert. "The Guadalupe Fur Seal: Habitat, Behavior,

Population, Size and Field Identification." *Journal of Mammalogy,* vol. 49, no. 4, pp. 665–675.

Scheffer, Victor B., *The Year of the Seal.* New York: Scribner's, 1970.

————, and Kenyon, Karl W. "The Fur Seal Herd Comes of Age." *National Geographic Magazine,* vol. 101, no. 4 (April, 1952).

Seed, Alice et al. *Seals, Sea Lions and Walruses.* Seattle: Pacific Search Books, 1972.

SEA TURTLES

Booth, Julie, and Peters, James A. "Behavioural Studies of the Green Turtle, *Chelonia mydas,* in the Sea." *Animal Behaviour,* vol. 20, no. 4.

Carr, Archie. *Handbook of Turtles.* Ithaca, N.Y.: Comstock Publishing Association, Cornell University Press, 1952.

Ernst, Carl, and Barbour, Roger W. *Turtles of the United States.* Lexington: University Press of Kentucky, 1973.

Fehring, William K. "Discrimination in Hatchling Loggerhead Turtles (*Caretta caretta caretta*)." *Animal Behaviour,* vol. 20, no. 4, p. 632.

Pritchard, Peter C. H. *Living Turtles of the World.* Neptune City, N.J.: T.F.H. Publications, 1967.

WHALES

Alpers, Antony, *Dolphins: The Myth and the Mammal.* Boston, Mass.: Houghton Mifflin, 1961.

Burton, Robert. *The Life and Death of Whales.* New York: Universe Books, 1973.

Ciampi, Elgin. *Those Other People, Porpoises.* New York: Grosset and Dunlap, 1972.

Cousteau, Jacques-Yves, *The Whale—Mighty Monarch of the Sea.* Garden City, N.Y.: Doubleday, 1972.

Devine, Eleanor, and Clark, Martha. *The Dolphin Smile: 29 Centuries of Dolphin Lore.* New York: Macmillan, 1967.

Flower, William Henry. "On the Recent Ziphoid Whales, with a Description of the Skeleton of *Berardius arnouxii.*" *Trans-*

action of the Zoological Society of London, vol. 8, pt. 3 (November 7, 1871).

Gardner, Erle Stanley. *Hunting the Desert Whale.* Toronto, Ontario, Canada: George McLeod, Ltd., 1960.

Kellogg, Winthrop. *Porpoises and Sonar.* Chicago: University of Chicago Press, 1961.

Nemoto, Takahisa. *Food of Baleen Whales in the Northern Pacific.* Tokyo, Japan: Scientific Reports of the Whale Research Institute, page 33, 1957.

Nishiwaki, Masaharu. *On the Sexual Maturity of the Sperm Whale* (Physeter catodon) *Found in the North Pacific.* Tokyo, Japan: Scientific Reports of the Whale Research Institute, p. 39, 1956.

Norris, Kenneth S. et al. *Whales, Dolphins and Porpoises.* Berkeley: University of California Press, 1966.

Ommanney, F. D. *Lost Leviathan.* New York: Dodd, Mead and Co., 1971.

Omura, Hideo; Kazuo Fujino; and Seiji Kimura. *Beaked Whale, Berardius bairdi, of Japan, with Notes on Ziphius cavirostrus.* Tokyo: Scientific Reports of the Whale Research Institute (June 1955).

Omura, Hideo, and Haruyuki Sakiura. *Studies on the Little Piked Whale from the Coast of Japan.* Tokyo, Japan: Scientific Reports of the Whale Research Institute, no. 11 (June 1956).

Lilly, John C. *Man and Dolphin: Adventures on a New Scientific Frontier.* Garden City, N.Y.: Doubleday, 1961.

Rice, Dale W., and Allen A. Welman. *The Life History and Ecology of the Gray Whale.* Special Publication No. 3, The American Society of Mammalogists (April 1971).

Riedman, Sarah R., and Elton T. Gustafson. *Home Is the Sea for Whales.* Chicago: Rand McNally, 1966.

Scheffer, Victor B. *The Year of the Whale.* New York: Scribner's, 1969.

Seed, Alice. *Baleen Whales in Eastern North Pacific Waters.* Seattle, Washington: Pacific Search Books, 1972.

Slipjer, E. J. *Whales.* Trans. from the Dutch by A. J. Pomerans. New York: Basic Books, 1962.

Small, George L. *The Blue Whale.* New York: Columbia University Press, 1971.

Uda, Michitaka, and Keiji Nasu. *Studies of the Whaling Grounds in the Northern Sea Region of the Pacific Ocean in Relation to Meteorological and Oceanographic Conditions.* Tokyo, Japan: Scientific Reports of the Whale Research Institute, no. 11 (June 1956).

Watkins, William A. "Air-borne Sounds of the Humpback Whale, *Megaptera novaengliae.*" *Journal of Mammalogy,* vol. 48, no. 4.